CB BIBLE

CB

BIBLE

By Porter Bibb

with Peter Livingston and
Michael Marcus

Designed by Seymour Chwast
Photographs by Jean-Pierre Laffont
Drawings by Bob Neubecker and Susan Willmarth
CB Comics by Dick Seigel and John Butterfield
Produced by Jean-Claude Suares

Dolphin Books
Doubleday & Company, Inc.
Garden City, New York
1976

Credits

pg. 5, Smokey Enterprises, Inc.; pg. 11, © 1976, Bill Fries; pg. 12/13, Patches by A.T. Paradice and S. Russell; pg. 22, Broadcasting magazine; pg. 30, Malcolm Lea, S-9 magazine; pg. 32, "Movin' On"-NBC-TV; pg. 33, S-9 magazine; pg. 38/39, Dick Bleil; Coffee Break News, Two Rivers CB Club, S-9 magazine, CB News & Truckers Gazette, Paper Doll; pg. 40/41, CB News & Truckers Gazette, S-9 magazine, Dick Bleil, Toad Frog; 49/50, John Butterfield; 51/53, Dick Siegel; pg. 55, Overdrive magazine; pg. 60, Mike Blesse pg. 61, REACT; pg. 62, Kathy Gurley; pg. 70, Donna Price; pg. 74, Broadcasting magazine; pg. 123, Charles Del Vecchio, Washington Post; 129, S-9 magazine; 173/175, Michael Marcus; 183, S-9 magazine; Special thanks to Tom Kneitel and the staff of S-9 magazine.

PUSH PIN PRESS

Producer: Jean Claude Suares
Editor: William E. Maloney
Design Director: Seymour Chwast
Copy Editor: Leslie Holzer

Library of Congress Card Number 76-11287
ISBN: 0-385-12323-X
Printed in United States of America
 First Printing
Dolphin Books Original Edition 1976

This book is dedicated to
CBers Everywhere, but most
especially to "Alice Rimple Dimple"
and "The Flower."

10-4

BREAKER
CB
COME
ON

Table of Contents

Foreword

For a long time people have been looking for new ways to talk with strangers and meet people, and CB is it. You can talk to somebody you have never seen, discover that you have something in common, and never have to get caught up in somebody else's life. Then, there's a certain magic about "10-4" and radio talk in general, a kind of mystique that I'm sure is partly responsible for the unbelievable growth in CB popularity.

Of course, a whole big part of the CB world is the language, and if you want to get into CB, you have to learn it. It's truckers' talk and ever since I was a kid and used to ride along on the big flat semis that carried the prefabricated farm buildings my dad sold, I've been fascinated with trucks and truckers. They're a kind of foot-loose, devil-may-care cowboy, and they lead a life style that appeals to the free spirit in us all. They go where they want to, when they want to, and mostly they're their own boss. CB is a big part of their world. And it's an exciting and romantic place to be. You never know what's going to happen next.

Actually, though, I bought my CB for more serious reasons. I wanted to have some sort of communication available in my jeep when I'm out in the boonies. CB is definitely the best safety device since seat belts and anyone who does much driving should have one.

As you'll soon discover, CB is like a passport to a whole new collection of experiences. And no matter where you go, whether you're in the city, on the highway, or out in the wilds, you'll find someone to talk to and something to talk about. That's the beauty of citizens' band. It's as good as it's name and just about anybody can get on the air with a minimum of hassle.

I'll have my ears on and one day this "Rubber Duck" may even copy your mail. In the meantime, the air waves are yours and you'll find yourselves modulating with cottonpickers you never would meet any other way. Now get out there and start blowing smoke. You've got a bumper crop of those Big Numbers from ole C.W. The rest is up to you.

I'm down and on the side.
C.W. McCall

C. W. McCall (Bill Fries) is a songwriter, singer and former advertising executive. His CB hit, "Convoy," burst across the country and western charts and, in early 1976, onto the top of pop music Hit Parade. A musical fantasy of a thousand truckers who use citizens band to organize a coast-to-coast convoy, McCall's great hit helped popularize CB throughout the country and around the world.

Introduction

Owning and operating your own radio station was never a big part of the American Dream. But today, any man, woman, or child (in his parent's name, if he's under 18) with $100 and a strong desire to get on the air can become part of a nationwide "network" of over ten million citizens band broadcasters.

Limited for years to a small universe of amateur radio buffs and over-the-road truckdrivers, CB has suddenly taken on the proportions of a full-fledged cultural phenomenom. Everybody is sprouting ears (CB antennas), motorists, police, firemen, hunters, boaters, campers, construction workers, doctors, farmers, housewives, and just plain CB nuts. The air is clogged with ratchet jaws (nonstop talkers) and rubberbanders (neophyte CBers). CB songs have infiltrated the Hit Parade and CB talk is, almost unnoticed, working its way into our everyday language. The business of making and selling CB equipment has already become a multibillion dollar industry, all in little more than 24 months!

What exactly is citizens band radio? The Federal Communications Commission (FCC) in 1958 set aside a group of high-frequency radio channels exclusively for short-range, two-way voice communications. These channels, which are in the 11-meter band, have become known as Class-D CB. A station license is required by the FCC before you can transmit on CB, but it's no more difficult to obtain than a fishing license. No examination or technical knowledge is required and anyone over 17 can get a license by paying $4 for a 5-year ticket.

How does CB work? CB transceivers are light-weight, low-powered, relatively inexpensive two-way radio sets. Each transceiver unit contains a transmitter and a receiver, and a power supply for providing proper power to both. Transceivers are available that operate on either 12 volts DC (eg, a car), 120 volts AC, or both. Antennas may be mounted on the transceiver itself, or via cable, at another point.

There is a wide choice of available CB equipment, everything from hand-held, battery-operated walkie talkies to mobile units for all types of vehicles, including cars and trucks, boats, planes, even motorcycles and snowmobiles. Both single and multichannel transceivers are available. Single-channel units can send and receive only from other units operating on the same frequency. Multichannel units offer over 50 different frequencies.

The reliable range from a base station to a mobile unit on flat or gently rolling country is about 20 miles, although a base antenna located on a tall building or on a hill, can increase broadcast range to perhaps 30 miles. Over-water range from boat-to-boat or boat-to-shore varies from 10 to 25 miles or more, providing proper antennas are used and no obstructions block the sightline.

CB signals are sometimes reflected back to earth from layers of the ionosphere and may produce moderately strong signals at distances from several hundred to several thousand miles. This phenomenon is called "skip" and usually occurs in winter and is most prevalent during certain years in which sunspot activity is greatest. The maximum legal range of CB transmission set by the FCC is 150 miles. The principal restrictions governing normal, licensed citizens band use are rules against profanity and advertising.

In spite of rapid advances in the state of the art, CB technology has had little thus far to do with the surging popularity of CB radio. The basic transceiver remains relatively uncomplicated (although the FCC prohibits anyone except a licensed service man from making critical adjustments on the set). The real emphasis has always been on communicating. And the CBer communicates with a vengeance. Through CB he's talking over the backyard fence again; he's back on the old telephone party line; he may even be eye-balling his new neighbor at an old fashioned "coffee break." Since the FCC has made CB virtually regulation free, he can turn CB's democratic communications potential to any use he cares to make of it.

Cloaked in the anonymity of his handle (CB nickname) the CBer is a fiercely independent, if frankly amateur broadcaster. He has become part of an invisible nation of the airwaves. And his numbers are swelling by the millions every year. In an age where conversation has given way to the color tube, CB has got people talking to one another again. And anything that can do that for America in the mid-1970s is nothing short of fantastic!

Porter Bibb,
Wheeling, West Virginia
April, 1976

1.

Being an Inquiry, into the Social, Economic, and Political Causes and Nature of the Phenomenon Known as Citizens Band Radio

On a wintry March morning in 1894, Thomas A. Edison sent an assistant scambling along the roof of a railroad caboose to make a final check of the connectors on the primitive loop antenna he had earlier installed. Inside the caboose, another assistant stood ready at a telegraph key. A long piece of exposed copper wire ran along the ground, parallel to the tracks, a few feet from the caboose. One end of the wire was connected to a telegraph "base station." The other end, apparently, just ran out down the tracks and wasn't connected to anything. As a switch-engine stoked up and began edging the caboose through the switchyard, Edison pounded out a series of dots and dashes on his base-station telegraph key. An inductive field around the transmission line was "sensed" by the loop antenna on the caboose and carried inside, causing the telegraph sounder there to begin clicking. Edison then signaled for his man in the caboose to reverse this procedure. Soon, the base station along the tracks was vibrating with his assistant's response.

Although the distance between Edison's copper wire transmission line and the loop antenna on top of the caboose was only a few feet, he had created the first system of mobile electronic communication. The same basic technique would still be in use 50 years later by railroads utilizing low-frequency radio-telephone transmitter-receivers and would lead, ultimately, to a universal citizens' radio system. But Edison had achieved his accomplishment six years before Nicola Tesla had even invented radio and over half a century was to pass before his original discovery would be fully realized.

Today, two-way, wireless-radio communications has become as commonplace and, for many Americans, almost as indispensable as the telephone—the wired communications system Edison had set out to improve upon. But it took enormous technological developments during the first four decades of the twentieth century before Edison's concept of personal, two-way wireless communications for the average citizen could become a practical reality. Thomas Edison had a powerful acolyte, however, in the person of Ewell Kirt (Jack) Jett, a former chief engineer of the FCC who began advocating, during the early 1940s, a system for widespread personal communications. Jack Jett was awed by the great technological improvements WWII had brought to radio and he pressed, in early 1945, for the establishment of a Radio Technical Planning Board to explore the possiblities this new technology offered in providing radio communications for the average citizen. Composed of FCC and other government officials, electronics manufacturers, and various interested communications specialists, the Board probed the potential of new radio technology and suggested that new frequencies be set aside exclusively for citizens' use. Acting on a Board proposal put forward by Jack Jett and radio pioneer Dr. Daniel E. Nobel, the FCC on December 1, 1947, created the Citizens Radio Service and made available for the first time the long-awaited convenience of a two-way radio-telephone system which could be now used by practically anyone.

Limited initially to one class (Class A) and assigned to the 460-470 MHz wavelengths in the ultra-high frequency region of the radio spectrum, this new service nevertheless enjoyed a gratifying, if hardly overwhelming reception from the general public. With a minimum of technical knowhow and a bit of rewiring, it was possible to put together a home-made, two-way citizens service transmitter with WWII surplus airborne transponders, tuned lines (antennas), and "doorknob" tubes. All this would cost only about $16. Even then, however, equipment manufacturers realized the great market potential for personal two-way radio communications and most of them set out to seduce the early users of the citizens radio with hardware that was more elaborate (and considerably more expensive) than those old home-made "breadboards." Old-time radio buffs remember with affection the Stewart-Warner Electric Company's Model 73 Portafone — a "hotline" type which looked like a regular telephone hand set with a 12-inch dipole antenna protruding from its base. This "citizens radio-telephone system" cost $199 per unit, but bringing any two units together on the 460 – 470 MHz frequency was a near impossibility, since crystal-controlled, locked tuning had not yet been perfected. The manufacturer claimed that the Portafone transmitted "many miles along unobstructed lines of sight" although under less favorable conditions, "the range will exceed only a few hundred yards." Stewart-Warner soon scrapped its citizens' radio project, but the handsome red Portafone has since become as much of a collector's item as are the original FCC-issued Class-A call letters, which all begin with "G" (Since 1958, all CB call letters have begun with "K").

Class-A Citizens' Radio was an immediate, if limited success. Overcoming the technical difficulties encountered within the 460 –470 MHz frequencies and staying within the FCC's rigid Class-A regulations meant using equipment that was usually too costly and/or too complex for the average citizen.

Hoping to save the operator money, the FCC soon introduced a Class-B Citizens Service at 465 MHz. Limited to a maximum of 5-watts output (against Class-A's 50 watts), Class-B equipment should have been much more reasonably priced and easier to operate. The limitations of the 465 MHz frequency, however, created unforeseen technical difficulties in tuning and reception —in spite of the introduction of rapidly improving hardware by many manufacturers. And a Class-B, 5-watt transmitter could still cost over $500). After establishing a Class-C Citizens Service—strictly for remote-control systems like model airplanes and garage doors — the FCC was impressed enough with the technological developments and widespread potential applications of Class-B equipment to consider relocating most CB operations to another frequency — 27 MHz. Jack Jett's dream of a cheap, effective, and simple system of two-way communications for every citizen was about to become a reality.

During the summer of 1958, the FCC took a long hard look at the difficulties surrounding their Citizens' Radio Service and decided that the 11-meter band, then the home of tens of thousands of dedicated "ham" operators — some of whom had even organized into a "Save 11 Movement" — could be put to better use by the millions of potential CBers. Accordingly, on July 11, 1958, an extensive revision of the Citizens' Radio Service rules was announced (together with a less than cataclysmic change in the Amateur "ham" Service), and the 11-meter band was taken away from the hams and reassigned to the newly created Class-D Citizens' Radio Service.

On September 11, 1958, the amateur radio band was officially reopened for communications as the new Class-D (23 channel) Citizens Band. Since that day, CB's official anniversary, the Citizens' Radio Service has experienced astonishing growth and has precipitated unprecedented technical progress in the design of two-way radio transceivers. There have been incalculable social and cultural innovations, including what amounts to virtually a new CB language, music, and way of life for millions of Americans and others around the world. In the late 1950s, however, CB was still an unexplored frontier and more than two-dozen American manufacturers rushed into the market, hoping to stake significant claims among the millions of potential CBers. Some of the great names in CB hardware (Johnson, Lafayette, Shakespeare, Hy-Gain) were among the earliest entries, but the very first Class-D transceiver on the market was the Kaar IMP. (Manufacturers of some of the finest equipment of that era, Kaar later bowed out when their over-engineered products became too expensive to be competitive.) Other early CB base units included the Heathkit "Benton Harbor Lunchbox," and the RCA RadioPhone. The first portable CB transceiver was made in 1959 by Vocaline, another quality electronics firm that gave eventully way to price-cutting competitors. The only similarity between these heavy, early tube-type, super-regenerative rigs of 1959 and 1960, however, and the small solid-state units of today is the fact that they both operate in the 27 MHz band. And most of those early rigs were priced out of the reach of the average consumer. The Kaar IMP, for example, sold for $360 in 1960; the Vocaline portable transceiver was over $500.

The original Kaar IMP.

The first FCC rules governing Class-D Citizens' Band use called for all licenses to make provision to receive CONELRAD Radio Alert and the CONELRAD Radio All Clear. This meant that every CB operator had to have immediate access to an AM or FM radio or a television set or a special CONELRAD receiver. In those days of "Red scares" and anticipated overnight nuclear attacks, CB operators were prohibited from using their call signs or giving their locations during a CONELRAD alert, lest the "enemy" find it easier to make a navigational fix. Those, days, however, soon ended, and never again would the FCC provide any legal excuse for operators to fail to identify themselves.

By the mid-1960s, over 1 million Class-D license applications had been received by the FCC. Since the applicant had merely to be a U.S. citizen, 18-years of age or older, virtually anyone was eligible. At first, the rush to Class-D, consisted mainly of radio operators who missed the 11-meter band and the new Citizens' Radio Service became in reality a kind of test-free ham band. When one popular electronics monthly asked, "Who will be the first to issue a certificate for 'Worked All States-Citizens Band?' " the door was opened for a loop-hole that amateur "DXers" (long-distance ham operators) found in the Class-D regulations. In spite of the fact that super-regenerative receivers were standard equipment and the transmitters gave out less than 5 watts, this was more than enough to work all the way across the country from coast-to-coast, relaying signals from station-to-station. At first the FCC looked the other way, then it became painfully obvious that the hams had taken over the new "business and personal" service — they were reclaiming the 11-meter frequency they had lost in 1958. Many manufacturers abetted this takeover by dropping ham equipment from their lines and switching to CB. With the sudden rush of ham, and pseudo-ham operators to 27 MHz, many little electronics equipment makers suddenly found themselves part of a burgeoning, multimillion dollar CB equipment industry. The American Citizens' Band Association, a quasi-lobby group, came into existence in 1962, and added its voice to the clamor for less FCC regulation, more power for CB transceivers, and very nearly total anarchy on the airwaves. The FCC took a firm stand, however announced several severe proposed changes in the rules governing Class-D Citizens service, and requested comments from the general public.

Jack Jett had predicted this situation when he wrote back in 1945 that "citizen's radio could be abolished . . . if its operation were not in the public interest, convenience, or necessity." The sham ham interlopers on Class D held their breath and waited for the FCC to act. Months dragged while no action was taken on the new rules. Meanwhile, the FCC stepped up its enforcement of rules violations (principally skip transmitting; operating without a license; and beefing up the 5-watt Class-D transmitter with a linear amplifier (power booster.) Simultaneously, legitimate users began to discover CB. Businesses, principally delivery fleets and independent and over-the-road operators, were quick to appreciate the enormous benefits an inexpensive, two-way communications system could be to an over-the-road driver. Other CBers were quietly setting out to improve the level of their operation, to organize for public service and emergency operations, and to achieve the goal for which the Citizens' Radio Service had been intended. And the FCC, affecting a characteristic *laissez-faire* attitude, was pleased to see that the service could, in fact, develop its own proper operating techniques and methods which the Commission might then translate into Class-D rules and regulations.

The CB boom had not quite begun, but by the mid-1960s, over one million Class-D license applications had been received and

Citizens Band Goes International

processed by the FCC and news of CB's accessibility was spreading. North of the border, the Canadian Government's Department of Transport initiated plans to create a parallel citizens' General Radio Service and although the land rush of new users is just beginning there, Canadians have already found CB indispensible in logging, trucking, ski patrols, and highway emergency assistance. Over 2000 walkie talkies were used for crowd control, security, and committee communications at the 1976 Summer Olympics in Montreal. And although many European countries discourage open use of the airwaves by private citizens, talk on the 27 MHz band is becoming increasingly heavy in France, Sweden, Belgium, and Italy. Anything approaching a Citizens' Radio Service is presently illegal in both the United Kingdom and the Netherlands and, though legal, it is somewhat restricted in Germany, due to power limitations (2-watt maximum output) and strong restrictions on antenna erection. A Swiss firm, Zodiac, is the leading distributor of (principally Japanese) CB transceivers throughout Europe. A French manufacturer, SECRE, produces a 50-milliwatt FM transceiver for the 27 MHz frequency. These sets, although priced considerably higher than their U.S. counterparts, have created a strong demand among consumers, who have recently begun to appreciate both the business and personal communications potential of CB radio. (France is one of the few European countries granting private licenses, providing some "professional" application of CB is included in the license request.) The French PTT recently opened up 12 channels for such "professional" users, although the maximum legal transceiver capacity in France is limited to only six channels.

During the late 1960s and early 1970s, the typical Citizens' Radio broadcaster gradually became more akin to the telephone user than to the ham-radio operator. The primary motivation stimulating this new use of CB was the transceivers' communications capability. Many new CBers, in fact, began to view their Citizens' Band radios as ordinary household appliances, and were uninterested in the technology or esotericism of ham-type radio broadcasting.

A colorful, creative language grew up around the truckers who used CB on the open road and this language, liberally laced with law enforcement codes and other technical jargon, became the official CB talk and soon separated CBers from their shortwave confreres. AM all-night disc jockeys picked up the new language from the trucking and CB songs they played and they then began to spread the word over their own 50,000-watt clear-channel stations which the truckers and other all-night travelers tuned to religiously. Large corporations like Hallicrafters and General Motors began to provide funds for Citizens' Band public assistance organizations (the largest, REACT, has over 2500 emergency assistance teams throughout the country and answered over 20 million distress calls in 1975). Local and regional CB clubs sprang up overnight and coffee breaks became a fixture on every ardent CBer's calendar. The big, once-a-year CB jamborees came later, spinning off from such business/social extravaganzas as WWVA's annual Truckers Convention in Wheeling, West Virginia. The invisible nation of CB broadcasting was beginning to take shape. As the 1970s began, CB had not yet quite come of age. But by early 1976, it could boast nearly seven million licensed (and unlicensed) operators.

Richard Horner, president of E.F. Johnson Company, a leading manufacturer of CB equipment, thinks the FCC's decision in 1970 "to reserve Channel 9 for highway emergencies provided the fuse for an impending CB explosion." Suddenly, ordinary motorists, as well as truckers all over the country, became aware of the tremendous public service potential of CB communications. "The temperature was further elevated," Horner says, "by economic considerations." The affluence of the 1960s carried over into the early years of the next decade and with it brought Americans out onto the highways in unprecedented numbers. Campers and other recreational vehicles sprouted like crab grass in the center strips of the nation's interstates; pleasure boats spawned like tadpoles along America's rivers, lakes, and streams. The public was in a buying mood, and for a hundred dollars or so, anyone could purchase the convenience of constant two-way wireless communications. It seemed like a bargain at the price.

Then, in late 1974, came the Arab Oil Embargo and suddenly fuel-conscious America was faced with a nationwide 55 mph speed limit. At the same time, squeezed by fuel shortages, a steady spiral of inflating operating expenses, and a wage-price freeze imposed upon them by the federal government, the nation's independent truckers staged a cross-country shutdown. Network newscasts showed hundreds of massed semi-trailers jammed wall to wall along major interstate highways, blocking important bridge accesses, intersections, and toll booths. Organizing this spontaneous, but impressively effective truckers' strike would have been impossible without the trucker's toy and joy, his little ole CB radio, which provided the necessary link among the literally tens of thousands of otherwise unrelated and unaffiliated gear jammers. And that fact was not lost on the network television audience. Overnight, awareness of the Citizens Radio Service leaped from somewhere around 5% to well over 50% of all Americans. Suddenly more and more people found themselves acquainted with somebody who had a CB or who was just about to get one. The 18-wheelers had unsnarled their shutdown and once again traffic was moving freely along the interstates. And it didn't take an expert to realize that, somehow, magically, those same CB-equipped truckers (not to mention the increasing thousands of ordinary motorists who were joining the brotherhood of CBers) hardly ever seemed to be picked off by the highway patrol's new radar units. Those truckers and their 4-wheel followers (who usually sported six foot whip antennas on their back bumpers, instead of the twin, mirror-mounted 46-inchers found on most trucks) seemed immune to the oppres-

The classic 1930s Astatic.

sive new 55 mph limit. Soon, people were hearing about a sinister character called "Smokey," and as CB songs like Kenny Price's "Let's Truck Together," C.W. McCall's "Convoy," and Cledus Maggards's "The White Knight" began the short crossover from the Country Music charts to the top of the Pop Hit Parade, the general public began to get the picture and the land rush, the CB explosion, the boom old Jack Jett had predicted 30-years ago, was finally about to happen.

In the face of the worst economic recession since the Great Depression (and in spite of the fact that auto sales were at a 25-year low), FCC license applications for Class-D permits in 1974 were 96% greater than 1973 applications. In mid-1975, the FCC created several long-waited rule changes which greatly simplified regulations governing Class-D service and simultaneously lowered the license fee from $20 to $4. Universal two-way radio communication was just around the corner. And with it, an

Jack Jett, CB visionary.

almost unimaginable rush of new users into CB land. In September, 1975, for example, over 200,000 new license applications were received by the FCC. In November, the number climbed to 300,000. By December, over 400,000 applications were received. In January, 1976, the total rose — to over 500,000! Charles A. Higginbotham, chief of the FCC's Safety and Special Services Division, which includes CB activities, expects the 1976 total number of license applications to be nearly four million. And according to Higginbotham: "at least another million CB operators will be modulating illegally — either because they haven't applied for a license or simply haven't received theirs yet, due to the unprecedented number of applications which have simply swamped our facilities."

CB radio has, indeed, come of age. Eight million transceivers were shipped last year, according to John Sodolski of the Electronic Industries Association, and the total will exceed ten million in 1976. When CB equipment and accessories are added together, they accounted for over $1.5 billion in sales last year. This year that figure could top $2 billion. These numbers become staggering when one realizes that the entire record industry accounts for just about $1 billion in sales. Sodolski and other industry experts see a continued growth of the CB market for at least another four or five years and then a leveling out as buyers begin to upgrade and replace their older equipment. But Detroit is already giving serious consideration to the installation of CB transceivers as factory equipment on all trucks, recreational vehicles, and passenger cars and that market alone will consume an annual total of some 8 million transceivers per year.

It took Jack Jett's dream of universal citizens' radio nearly ten years to become a possibility. Then, in the short space of only ten months, millions of Americans had suddenly became a part of his communications utopia. Today, there are nearly 18 million CB radios in use throughout the country. One out of every eighteen American families (one out of every nine farm families) now uses the Citizens' Radio Service. One out of every fifteen automobiles on the road is equipped with a CB transceiver, four out of every five long-haul trucks find CB indispensible. In addition, one out of every six pleasure boats sports a CB radio (in spite of the fact that the Coast Guard recognizes UHF/FM as the only official marine communications medium).

Jack Jett and his colleagues on the FCC had hoped CB would eventually supplement the telephone as an effective and economic means of two-way communications. Thomas Edison had been driven by much the same desire — "to unhook his words from the telephone's connecting wires." None of these early innovators in wireless two-way communications could have possibly imagined what an effective complement to the telephone CB radio would actually become. Nobody has thought to ask Ma Bell what she thinks of the situation but CB radio is rapidly moving into the top rank of (as Cledus Maggard puts it in his CB hit, "The White Knight"), "those four most important things in life."

2.

CB Culture: Social Notes on the "Invisible Nation"

Gear Jammers and Grease Spots: Origins of the Trucker Mythology And How CB Became a Part of It

Today's trucker is part of that proud American tradition of fierce independence which stretches back to the dusty frontier highways where stage-coach drivers, wagon masters, and cattle drovers fought their way across the continent. The modern trucker, shifting through his twenty gears, shares many of the same experiences of the wagon master who managed a team of twenty horses. The twentieth century trucker's log book and the old wagon master's diary both described a treacherous journey across hostile country. The man who rode "shotgun" on the stage coach also served as codriver, and the term "shotgun driver" was still very much in use during the 1930s and 1940s when hijackers (still a threat to truckers) were commonplace. Then, being a trucker meant you were responsible, not only for your load, but for your truck as well. Repair shops were unknown and once a vehicle rolled out of the factory, it was the driver's job to take care of it. Truckers prided themselves in putting a million or more miles on their rigs and having made every repair themselves. Old-timers don't think much of modern truckers who "bring their trucks in to have the ashtrays emptied."

The life of the long-haul truckers was as lonely as a stretch of desert highway on a moonless midnight. And long after seat cushions replaced the boards a driver once had to sit on—long after windshield wipers replaced the glaze that truckers put on their glass with a raw potato and saltpeter — long after the unpaved gravel roads had become two-, then four-, then six-, and now eight-lane interstates, the trucker still had only himself, his cargo, and his thoughts to contend with as he drove on through the never-ending stretch of oil-stained roadway. Only the sounds of his eighteen wheels singing along the concrete slabs broke the monotony of the road. Nevertheless a man could get in a lot of thinking during those ten or twelve hours on the road, and when he finally pulled off into a truck stop or diner, he was as hungry for good talk as he was for his dinner. He was anxious to hear about the road conditions up ahead, and about the location of police or the Interstate Commerce Commission (ICC) checkpoints. But what he hungered for most was someone to share his thoughts.

Nowadays, however, almost every long-haul truck sports a pair of 4-ft antennas. These silvery stalks are hooked up to a CB that rides high overhead, hanging from the inside roof of a trucker's cab or occupying the place of honor among all the dials and guages along his dashboard. Now, with a CB in virtually every truck tractor (and usually four or five auxillary CB speakers spaced throughout the interior so he can hear over the din of his diesel) the long haul trucker is in constant communion with his 18-wheeling friends, filling the emptiness that once characterized his monotonous journey with stories and valuable information. The old-fashioned headlight and hand signals (one blink meant "Hi"; two meant "slow down—danger"; a U-sign made with two fingers indicated police or radar were ahead) have given way to a rich radio vocabulary where each variety of law enforcement official has his own name. Reports of Smokies, Kojaks, county mounties, and green stamp collectors, down the road or "in the bushes" fill the airwaves.

The trucker, it turns out, would have invented CB if it hadn't been sitting there waiting for him to discover it. All those lonely years, those millions upon millions of lonely miles, with nothing but a few feet and a bit of steel and rubber keeping him from becoming another grease spot on the highway. But then, barely into the 1960s, he suddenly found he could walk into almost any truck stop along the interstate, fork over a couple of double-nickels, and come away with one of those new little Japanese toys. Soon the electronic myth-making process took over and the trucker began to experience himself mirrored a hundred times over throughout contemporary culture: on the all-night AM radio shows where Big Juicers (AM deejays) like Bobby Cole or Charlie Douglas programmed exclusively for a trucking audience; on the tube where a trucker could watch Claude Akins play trucker in the TV series "Moving On"; or on the Late Show's rerun version of Stephen Speilberg's classic trucker film, "Duel"; or he might catch himself on the silver screen of the local drive-in, where "White Line Fever" might be playing with "Truck Stop Woman" or "Moonfire" (a $500,000 extravaganza produced by *Overdrive*, the truckers' favorite magazine).

It wasn't always this way. Before the old six-cylinder horse was replaced by an "18-wheel office," the trucker had done a pretty good job of fashioning his own image—a cowboy with his country western music, his bouffant, blondined women, and his tight-lipped camaraderie. But as he rolled out greeting each new dawn on the unwinding, neverending interstates, he was already heading straight for change. And each day that passed confirmed once again that the old days were gone for good.

"Hobo Jack" knew those old days well. He remembers when the money first started to flow along the highways and with it the first hints of bureaucratic red tape. Jack, who left home when he was fourteen and signed on as a relief driver out of Pittsburgh after five years on the loading ramp of a steel mill, even remembers the 1935 Motor Carrier Act, which was passed to control interstate transport. Everybody was dead broke then, but the new ICC was enough to drive a stake through the heart of even a healthy trucking industry.

Jack says he lost an eye back in '37 in an accident outside of Youngstown, Ohio. His truck was rammed and burned by militant truckers who weren't allowed to drive into Pennsylvania. Nobody was ready for the backlash but the government had to find out "the hard way." Jack says, "The only time a trucker will show his nasty side is when he's made to stay in one place too long." Jack remembers the first sleeper cabs, introduced in the late 1930s by Kenworth; the first turbine engines, quietly discarded when the diesel was developed. He remembers when a trucker's wallet was called a "chain-drive pocketbook" because it had a habit of "driving off by itself" in truck stops and roadhouses. Hobo Jack remembers as if it were yesterday.

Hobo Jack lives on his Wisconsin "funny farm" today, within sight of Interstate 15. He has lived alone for the past 15 years (his wife died in 1961 and his two boys have long since moved away). The old 1958 K-Whopper in his driveway has logged over two million miles and won't likely see more than another few thousand or so.

Hobo Jack's only 60-years old, but he has the tired look and torn-down body of a man 20 years his senior. "Trucking does that to you," he comments knowingly, and a visitor can't help noticing the road maps of wrinkles that spread out from the deep sockets housing his eyes. He's healthy enough, says Jack, but what else besides a lifetime case of hemorroids and a ghostly grey pallor can 30 years in a truck give a man? He beckons you into the kitchen with his good eye, takes you past his new eight-foot Frigidaire and into the pantry. Suddenly, Hobo Jack is in another world, for it's all here in a room no more than 4-ft square, that he recaptures the life he led. Sitting down before an old E.F. Johnson Messenger II mobile-base combination, Jack opens up first one CB conversation, then another, and another with passing truckers. He can sit here six or eight hours at a stretch, almost as long as his old turns behind the wheel of his "Kan't-Work." And it's hard to pass through southern Wisconsin without hearing some old buddy of bygone days or a newfound CB buddy sending out a collect call for "Hobo Jack." Most of the midwestern truckers today know about Jack and many have even turned off the interstate to share a meal with the oldtimer. And once another 18-wheeler pulls up outside Jack's small roadside spread, the "One Hobo" knows he's in for an evening of reminiscing about a world he'll never see again. Jack's CB, though, came right out of his truck, bringing with it a million miles of memories. No driver who stops to share a meal or bunk down for the night ever leaves without an earful of trucker stories that will last him all the way to the West Coast and back. Jack was (and remains) a drinking man and his guests have seen him put away a fifth of Early Times faster than a diesel can tank up. But Jack wouldn't have so many guests if it weren't for his trusty Johnson and many a dawn will find him still modulating behind his mike as the morning light stretches across his little farm and onto the interstates where his gear-jamming brothers still make the long haul.

"Dracula" whose home 20 is Cleveland, knows Hobo Jack well. And most of the drivers who pass through Ohio, Indiana, Illinois, Missouri, Wisconsin, and Kansas know Dracula. A wizened, wirey little man who looks like he'd never be able to see over a steering wheel, Dracula should have been born 20 years sooner. If he had been, Hobo Jack the rest of the world would probably have known him as the Bela Lugosi of Commercial Radio. He has the Count Dracula character down to a bloodthirsty T, but it is confined to Channel 19 of the citizens' band, where he thinks nothing of carrying on a monologue for hours

(while everybody else on the "diesel digits" takes a back seat). He calls Pennsylvania "Translvania," naturally, and taunts the highway patrol with thickly accented threats: "The only vay you pipple vill cvatch me," he will spit out, "is with a silver speeding ticket through my hee-ort." Dracula seldom pulls those tickets, however, and is ironically on good terms with most Smokies, who know he'll go out of his way to report a drunken driver or assist at an accident.

Another open-hearted CB-trucker is "Rug Catcher," who's from Las Cruces, New Mexico. This unlikely fellow monitors Channel 20 and will meet you, so he says, any time day or night with a full supply of new and used transceivers, antennas, scanners and replacement parts, just in case your own CB rig has burned out.

The Rug Catcher stays around his home 20 and he's a 9-to-5er, a company man logging little more than 300–400 miles a day. His counterpart, the long-hauling, over-the-road driver, may be a wild-catting owner-operator, or a teamster driving a company truck, but either way, his truck is his home. And if he's one of the 100,000-odd independent truckers still left, then he truly qualifies as the "Last American Cowboy." He can make good money, maybe $18–20,000 a year after expenses and his truck payments are deducted, but he has his freedom and nobody but nobody can put a price on that. The long hauler may be married, but he'll see his wife only a week or so out of every month. He'll be lucky if he can remember his kids' ages, much less their birthdays, and he may even be maintaining more than

one family—each a coast apart. Nevertheless, he lives on credit, since his money will most likely go right back into his truck or time payments, "buying off the furniture."

"I sleep better in my truck than I do at home." That's The "Spiderman" talking, and he won't see his home in Bossier City, near Shreveport, Louisiana, more than twelve times a year. "Sometimes my wife will wake up and find me on the living room couch. She thinks I'm crazy but it feels more like my truck to me." The Spiderman is a twice-over bigamist; married to his wife, to his codriver, and to his truck — a big blue and white cab-over-engine Peterbilt. The Spiderman probably owes twice his annual income at any given time, either to his bank, his finance company, or his wife. But since he bought his first CB rig, about 10 years ago, The Spiderman has given away more money than he's made to handicapped truckers and a host of other needy causes. Putting his lilting southern accent to good use, Spiderman sends a constant stream of charity requests out over Channel 7. His fund-raising efforts stretch from the Gulf Coast of Florida all the way west to Corpus Christi, Texas. Truck stops throughout the south know about his efforts and put collection tins out on their counters marked with his name and the current charity recipient. A relatively young man of 31, the Spider looks 15-years older and it's only when a stranger probes a little, asking about his limp and the strange, stiff way he has of walking — too stiff, even for a trucker — that his motivation becomes clear. Almost 10 years ago, The Spiderman was carrying a load of cotton through the east Texas flatlands. His truck was rammed, head-on, in the middle of the night, by an Airstream housetrailer that had broken loose from its trailer hitch and was free-wheeling right in front of his windshield while he was doing over 70 mph! A freak accident, to be sure, and one that left Spider with a permanently shattered heel when they pulled him from the wreckage. The driver of the pick-up pulling the Airstream had tried to warn him, via CB, but the Spider hadn't got his ears yet. He became a confirmed CBer, as well as an obsessive fundraiser only when he learned that an 8-year-old girl had lost her life inside that Airstream.

"Poor Boy" trucks cigarettes up from North Carolina. He knows all about hijackers and has had his trailer "lifted" five times, the fifth time two years ago. The next day he went straight out to a truck stop outside of Trenton and bought himself a $250 Midland mobile unit, one he considered "damned near the best on the market." He had to buy a detachable-type antenna, because his trucking company didn't allow their drivers to have CB (many drivers still must buy their own rigs and operate them against company policy). Since installing his radio, Poor Boy hasn't even seen a hijacker, much less lost any loads to the looters. "One sight of those ears scares them away," he says confidently. "Having ears" also makes his time a lot more enjoyable when he has to wait a couple of hours at some warehouse waiting to be off-loaded at two or three in the morning.

There are more trucks in Texas than in any other state in the union. And Texans, used to spanning great distances just to get from one place to another, seem to fall naturally into the trucking life style. Everything is bigger in Texas, a Texan will tell you, including the roads. They're just about the widest and longest in the world. The largest load ever hauled by a truck, a Saturn rocket used on the first Apollo missile, however, was shipped out of Huntsville, Alabama. The truck that carried it was shrouded with white tarps. Dubbed "Moby Dick" by the driver, this 80 ft-long load was protected by a security force of eleven armed guards, seven police cars equipped with external P.A. speakers (who "announced" the caravan) and eleven other vehicles with danger signs and flashing lights which followed in its wake. "Moby Dick" traveled across the south to Cape Kennedy and the trailers tires were deflated seventeen times to gain the vital extra inches of clearance needed for this great white whale of a truck to clear overpasses and bridges.

"Bedbug Barney" has never carried a Saturn rocket, but he's a furniture hauler with an astonishing list of loads. He once hauled an atomic bomb from the Rocky Mountain (Colorado) Arsenal to Alamogordo, New Mexico. "They wouldn't let me watch while I was being loaded," Barney says wistfully, "and they had guards with machine guns all over the place." In August, 1974, Barney was asked to help move the Nixon family out of the White House. "I'd have done that for free," Barney says, "but they insisted on paying us. Six drivers and three truckers—each van was insured for over a million dollars. All sorts of nice stuff. Probably gifts from foreign governments." Barney and the other drivers used their CB radios to maintain contact in their three-truck convoy all the way to San Clemente. "There were plenty of people copying our mail all the way from Washington. Everybody wanted to know if Nixon was going to make a comeback. How the hell would we know? All I told them was that we had his furniture on board and that the stuff was on a one-way trip."

Truck driving had come of age in the 1950s. And by the 1960s, drivers who drove for the love of the road either had to learn new rules or quit the game. Gypsy drivers, those leathernecks who trucked without permits or authorized rigs, faced stiff penalties if caught. Truckers were becoming small businessmen and their trucks were their offices, and all too often they found themselves negotiating tricky turns around a mountain of paperwork. By the early 1970s, the trucking industry had become a $20 billion a year bonanza. But despite the fact that the government insisted on viewing him as a "common carrier," navigating his way through a thicket of weigh stations, log books, fuel tickets, and mileage records, most truckers continued to see themselves as knights of the road, truly the Last American Cowboys.

Ironically it was an international fuel crisis that forced the truckers out of this free-spirited, independent stance and caused them suddenly to confront the world, for a moment at least, with a collective force that was almost "corporate" in its organization and efficiency. And it was this unexpected unity that also brought CB to national attention.

Shortages, the shutdown, & Smokey

It had been coming for almost a year, and by the time the leaves began to fall in the autumn of 1973, more than a few truckers had already experienced the long lines at the fuel pumps that awaited all American motorists a few weeks down the road. Suddenly everybody was "getting ears" and CB was about to come into its own. Some 20,000 long-haul truckers jumped on the CB bandwagon during the last three months of the year. Then the Arab Oil Embargo became official. Nobody but the truckers seemed particularly upset. The 4-wheelers huffed and puffed, but still got their gas after a little waiting on line. The teamsters and company organizations pontificated. The White House was having its own difficulties and responded to the crisis with talk of fuel price hikes and speed limits.

Enter Mike Parkhurst, rampaging, crusading, evangelical editor of *Overdrive* magazine, and champion of the independent trucker. Working out of a radio/telephone command post in the magazine's Los Angeles offices, Parkhurst organized a dramatic, bold reply to the lethargy and confusion coming out of Washington.

The morning of January 31, 1974, broke crisp and cold across most of America. At first the reports filtering in on the morning traffic reports hardly seemed worth a second notice. "Rush hour tie-up on the Jersey turnpike . . . A mile long traffic jam outside of Chicago on I-75 . . . a larger than usual delay in commuter traffic coming into Los Angeles." Slowly, the picture became clear. There were no ordinary tie-ups; tens of thousands of the nation's million truckers had seen to that. Incensed at rising fuel costs, their increasing inability to get any fuel at all, and the sudden imposition of a nation-wide 55 mph speed limit, the truckers had simply rallied to Parkhurst's call, pulled their rigs into strategic positions blocking key interchanges, accesses and interstates all across the country, and then, incredibly, switched off their engines and walked away. And the whole staggering feat had been orchestrated by CB. By what? Suddenly John Chancellor and Walter Cronkite and everyone in the network news departments wanted to know about CB. What was it? How did it work? Were truckers the only ones who had it? The trucker, who had choked off the life line of the country, was suddenly given a chance to flex his muscles and rise above the myth of his slightly atavistic existence, to show his face, and speak his piece. Ultimately, however, the truckers' shutdown of 1974 was a failure. It did give the public a chance to scrutinize the trucker, and to see the men and women who carried the freight. But what the shutdown did not do was roll back the fuel prices or change the speed limit. Worse, many truckers, already mortgaged up to their eyeballs, had simply stayed out of the whole business, and suddenly found themselves squared off, billy-to-billy, sometimes

even gun muzzle-to-belly in a number of bloody confrontations that proved the trucker is nothing if he's not an independent cuss. Nobody, not even another trucker, tells him what to do.

Out of all that furor, what did happen was a rampaging awareness of CB among truckers, the public, and members of the nation's highway patrols. The new 55 mph speed limit made almost every motorist a criminal the minute he hit the accelerator. Worse, the new speed limit stood to put many truckers out of business. To meet their schedules, an independent over-the-road driver quickly discovered that CB was no longer a luxury—something to while away the lonely hours with. Citizens band had now become for him an indispensable necessity. Convoys, with "front" and "back doors" scouting speed traps, were organized spontaneously over CB. "Smokey reports," as well as indications of road and weather conditions up ahead, were passed via CB back and forth through the air above the median strip.

A whole new language (part trucker, part amateur radio operator, part police and military), sprang up as the truckers learned to speak in shorthand. They quickly took handles (nicknames) for themselves to avoid being identified by the state troopers they had just spotted. Before long, motorists began to fall in between the truckers' convoys. They weren't sure why yet, but one thing was certain—the trucks seemed to breeze along at 65, even 70 mph and then knew *exactly* when to bring their speeds down to the legal limit. Now those little antennas that had sprouted for several years from so many tractor cabs began to take on significance and 4-wheelers joined the rush to citizens' band.

Today there are well over 5 million mobile units cruising the highways in cars, trucks, campers, and house trailers. Both General Motors and Chrysler have already begun to install CBs as factory equipment, and industry sources estimate that the 5 million total will have doubled by the end of 1976.

Today, even Smokey has a good word to say about CB, in spite of the fact that many law enforcement authorities originally felt less than kindly toward this new electronic interloper. With over 41 million emergency and distress calls being carried over CB's Channel 9, countless lives and millions of dollars have been saved because of CB. Highway patrols in over a dozen states have added citizens' band units to their own existing communications systems. And most law enforcement officials now feel that drivers with CB are generally more alert, cautious, responsible and informed, and generally too busy passing along Smokey reports to bother speeding. Even The Federal Department of Transportation has begun promoting the use of CB in a citizen-police road safety network that has proved effective in several areas around the country.

And now what's happening to the truckers and their world on wheels? They're being overwhelmed in a rising wave of CB popularity; they're losing their once-exclusive hold on the truckers' channels (which are now as clogged with 4-wheeler chatter as they once were filled with trucking stories and Smokey reports). The Last American Cowboy, is now at the mercy of the media which is making him a folk hero before his time and yanking him into the 1970s by his ears.

Social Life of a Biscuit Burner

On April 12, 1975, the Lemon Bay Misfits of Englewood, Florida, together with several other newly organized CB clubs, sponsored the first "total CB" wedding. The lucky couple, Judy Bozie and James T. Wilson, were married in Englewood's rambling Indian Mound Park, with a CB justice of the peace presiding. All the participants (except the bride's and the groom's parents) were CBers. Which only goes to prove that while the truckers were discovering CB, plenty of nontrucking two-way communications addicts had been pretty busy developing a CB culture and society of their own. And as the worlds of trucking and CB merged in the mid-1970s, new life coursed through the veins of the old-line CB establishment, which, for nearly a decade, had been trying to recover from the character attacks heaped upon it by the high and mighty ham operator fraternity. The biscuit burners (from the earliest tube sets, which were called breadboards or biscuit burners) fortified with a new sense of pride, began to expand their social life. And as they did, they threw themselves into it with a vengeance.

Back in the mid-1960s, however, there were only a handful of statewide and regional CB clubs and organizations. In 1967, the first annual National CB Jamboree was staged at the Oklahoma City State Fairgrounds. Most of those who showed up for the two-day event found themselves choking on a nearly indigestible mixture of technical one-upsmanship (will your power mike outperform mine?) and tanking-up (on everything from Coors to California Ripple). The entire affair was tied together with a healthy dose of country, western, and trucking music. And in spite of the fact that fewer than 1000 people showed up, the pattern for future events was set. A coffee break in any of a dozen larger towns today will attract twice the crowds that rattled around the cement-block buildings of that first national get-together in Oklahoma City, and even regional jamborees like the one held every summer in Glens Falls, New York, will today consistently draw 10,000–15,000 avid CB addicts.

In early 1973 (months before the fuel crisis and subsequent truckers' shutdown), *Overdrive* held its first CB Pow Wow in conjunction with the magazine's own annual Trucker's Convention. For the first time ever, truckers, CBers, and two-way communications specialists (manufacturers, retailers, and distributors) sat down together in the same room to talk CB. Jerry Reese, national director of the newly reorganized REACT, was also there and introduced the participants to his growing Channel 9 Emergency Highway Assistance Program. (A year later, *Overdrive*'s CB Pow Wow would literally double in attendance.)

One of the oldest CB clubs in the country is the Charlotte County outfit in Punta Gorda, Florida. Joe Bicking organized

their 11th annual jamboree this spring. Jamborees and coffee breaks are usually part social, part CB business, part technical, and part public assistance. The Charlotte County CB Club, for example, has often raised money for needy causes, helped out in weather and traffic emergencies, and generally proved itself a strong asset to the local community. Most CB clubs, if they survive their first year of existence, prove to be focal points for popular community activities, especially in the South and Mid-West, where almost one in ten families depends upon CB communications almost as much as they do upon the telephone, AM/FM radio, or the television. Overcoming the anonymity of CB may be one reason so much socializing has grown up around CB — but perhaps the real reason thousands of state and local CB clubs sprang up so quickly is the genuine camaraderie that exists between most CBers. This spirit of shared adventure is almost inescapable the moment you first pick up a mike and send your first "Break " winging its way out along the invisible airwaves of CB land.

CB has always been a grass-roots movement but unlike truckers (who still resist any real attempts at organization), CBers have always grouped together when they felt the urge. While there were (and are) a number of so-called "national" CB organizations (including the American C.B. Association; the American Federation for CBers; the now defunct, but once notorious United CBers of America; and the recently organized C.B. Club of America), none really speaks for the biscuit burner nearly so effectively as his own local "coffee-break." Even their names have a homespun texture: Golden Mike Club (Aurora, Ohio), Crystal Busters Radio Club (Watertown, South Dakota), Hannibal Whips CBers (Fulton, New York), Jack Daniel CB Club (Tullahoma, Tennessee), CB Rangers (Butler, Pennsylvania), Cat Country CBers (Thief River, Minnesota).

The idea of clubs and coffee breaks grew out of the CBer's natural impulse (some would say perversity) to socialize. Then, when the FCC and the militant hams (who had given up the 11-meter band to CB in 1958) pressed for a little order on the airwaves, these occasional gatherings took on official overtones. Somebody had to speak up in defense of the CB, and in the absence of any effective national lobby, many of these scattered local and regional organizations slowly evolved into responsible forums where technical and performance information was exchanged and the latest FCC rule-change proposals debated. During the late 1960s, CBers gradually outgrew their defensiveness and began to see for themselves the dramatic possibilities for public assistance that CB offered. Many CBers aligned themselves with REACT or ALERT or one of the other independent emergency (Channel 9) monitoring services. Others participated in worthwhile civic and social activities, while still others explored the possibilities of playing policemen and established close links with local law enforcement officials. Intense pride swept CB land —club patches and bumper stickers blossomed, jackets, posters, and all manner of personalized geegaws and gimcracks were pawned off on giddy, but gullible CBers. "Official" handle registries materialized overnight to secure and protect (for $5-10) the exclusivity and individuality of carefully chosen CB nicknames. (No "register" can guarantee anything more than the fact that your handle will be printed in a national directory that might appear once a year—presumably if a dozen other CBers write in with the same handle, only the first one received will be printed. CBers, proud of their "original" handles, have taken to calling themselves "One French Fry" or "One Big Beard" rather than even fooling with the handle registries.)

QSL cards are another item that a lot of older biscuit burners are still caught up in. Derived from the ham's Q-code, QSL literally means, "Do you acknowledge me?" Shortwave transmitters are strong enough to carry a signal around the world, and hams have long acknowledged receipt of a strong distant signal by mailing the sender a QSL card. Over the years, these cards became the wallpaper of a proud ham's radio room and probably the only decor in his yak shack. CBers at first, eager to imitate hams, took up the practice. But since CB transmission is legally limited to 150 miles (although it is possible, of course, to skip many hundreds, even thousands of miles), the cards were handed out only at coffee breaks and other CB gatherings.

Several specialized CB publications, including two outstanding rival monthly magazines—*CB* and *S-9*—have been published for years, but recently have enjoyed skyrocketing circulation increases (*CB*, for example, jumped from 75,000 circulation in 1974 to over 200,000 in 1975; *S-9* also experienced a similar growth in readership). National CB newsletters, which had languished for years in the little-known (and less-read) backeddies of the CB mainstream, suddenly found themselves deluged with subscribers. At least two new CB newsletters for manufacturers began publishing early in 1975, and *Overdrive*, though hardly a general interest publication, has launched a CB section, which carries dozens of pages of CB information and advertisements each month, and has certainly become one of the best read and most reliable sources of CB developments anywhere.

Overdrive has even started its own CB club, Cross Country CBers, "for the serious CBer." Besides the usual membership cards, patches, and decals, new members receive a highway map of CB channels used throughout the United States, plus *Overdrive*'s support (which is not to be frowned upon), in case "your CB unit ever breaks down and the manufacturer hedges on the guarantee." Other publishers have also jumped on the rampaging CB bandwagon. Davis Publications, for example, which puts out the monthly *Elementary Electronics*, published a *CB Annual* in January, 1975. Sold out within two weeks, the *Annual* was immediately reprinted and another 125,000 sold just as quickly.

With the sudden surge of rubberbanders coming on the air, CB social life became a bit complicated. Serious activities like the

state conventions organized by the United States Citizens Radio Council, Inc. (P. Martin, President), had to compete with the likes of "Burro Control." Begun as a running joke between a few truckers who ran the lonesome roads of Nevada and Arizona, "BC" (CB, backwards) began putting out elaborate plans for monster coffee breaks, orgiastic jamborees, and a whole list of "bent" rules for new members. Somewhere along the way, Skip Streicher ("Marker Lite") and several others restructured Burro Control as a nonprofit organization, incorporated in Nevada, and dedicated to aiding handicapped children and other deserving people. Today there are Burro Control members throughout Nevada, Arizona, California, and seven other western states and a three-day "benefit break" is planned for next year at Beaty, Nevada, that should make CB history. Benefit coffee breaks and jamborees, as well as other fund raising efforts, in fact, are a big part of most CB club activities today. CB benefits may well become, if only because of their number, the telethons of the 1970s.

Clattering off the wires of the Associated Press, December 11, 1975, however, came one of the truest expressions ever of the spontaneous charitable camaraderie of citizens' banders anywhere:

(Lake Forest, Illinois)—Motorists who missed "Kit Kat" on their citizens band radio lately are back in touch with the woman truck driver . . . over a CB radio set up at her hospital bedside. Bobbie Jo Greene was forced off the road for six months when she broke both legs in an accident. During her hauls over highways of Indiana, Missouri, and Illinois, her citizens band call name was "Kit Kat." The 38-year old Miss Greene had not been in the Lake Forest, Illinois, hospital very long when her occupation became well known. A community project sort of sprouted to get a citizen's band radio for her hospital room, so Bobbie Jo could get back on the CB..

Ten-Four.

BINGHAMTON, N.Y.
CAMPING
AND CB'ING GO TOGETHER
HAND IN HAND
TWO RIVERS
CB RADIO CLUB inc.
BINGHAMTON, N.Y.

RIDE ON AIR
TWO RIVERS C B RADIO CLUB

GOLD COAST
NEBRASKA

WELCOME TO OKIE CIT

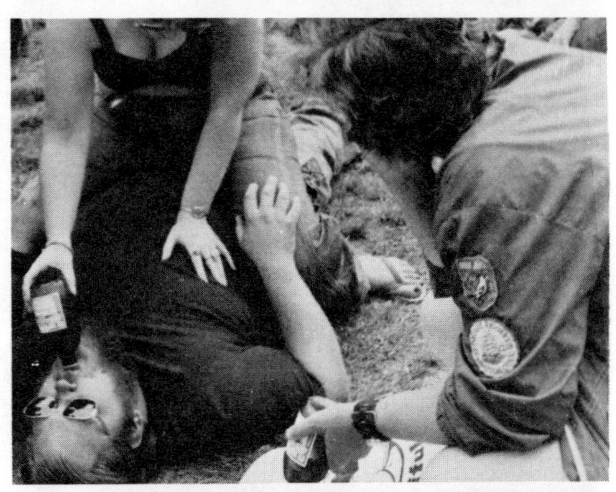

MEMBERSHIP
$5.50
P.O BOX
672
18 WHEELERS
CB CLUB

ORNERY MISPLACED OKIES
Subscribe now!
INTERNATIONAL
C B
NEWS
FORMERLY SOUTHLAND

PIONEER

Report from the CB Crime File

"Whatever kinds of problems people have in real life, they're going to have in connection with CB, because CB has become a real part of everyday life. It's a tool and that means if you're a criminal, an embezzler, prostitute, drug dealer, bookie, thief or kidnapper, you'll use whatever suits your purpose, including a CB radio."

James C. McKinney
FCC, Deputy Chief of Field Operations

"Break, 1-5. I've a collect call for the Red Dog, come'awn."
"Come-awn Breaker, you got the one Red Dog here."
"You got the Dairy Queen here outta A.T. Raleigh on U.S. 1. How are our bearded buddied doing on I-95?"
"Smokey's thick, good buddy. The cat's among the pigeons for sure."
"10-4, good buddy. You got an alternate route for me?"
"10-4, for sure. Take 1 to 85 into Petersburg town; then 95 to Richmond town and keep the pedal off the metal, looks like somebody spilled honey on the road, for sure. Them bears is crawling."
"10-4, good buddy. Hope to catch you on the flip-flop. Down and on the side."

Dairy Queen and Red Dog are out to avoid the bears for sure, but not because they're afraid of getting bit by a speeding ticket. Dairy Queen is driving a 1973 Chrysler and packed in the trunk and back seat are 1,500 cartons of cigarettes he legally bought from the Atlantic Tobacco Company (A.T. Raleigh) in North Carolina. He's a cigarette bootlegger and his 24-hour trip is worth $4,000. A carton of cigarettes is taxed $.20 when purchased in North Carolina; but in New York City, which is where he's heading, those same cigarettes would have cost $2.40 in additional taxes. It is, however, illegal to transport taxable goods across state lines for the purposes of avoiding sales taxes. Dairy Queen is small time compared to the trucks that make the same run with 10,000 cartons or more of cigarettes.

Cases like these, where CB is used for criminal purposes, are piling up at FCC headquarters. FCC Deputy Chief of Field Operations, James C. McKinney insists, however, that "while crime is increasing every year, the percentage of CB crime remains surprisingly low." The statistics indicate he's right, but with 500 agents and 100 assigned exclusively to CB crime, McKinney is finding the flood of new CB users (and accompanying increase in CB crimes) a difficult situation to deal with.

Armed with sophisticated directional finding (DF) equipment, his men located 4,000 violative CB operators and fined 3,000 of them in 1975 alone. In the same year his office issued 200 cease and desist orders to unlicensed CBers, revoked almost 300 licenses, and took 75 serious violators to trial. "And when we go to the trouble to take a man to trial," McKinney reports, "our case is usually strong enough for a conviction."

The more serious CB crimes, however, involve million dollar extortion attempts, large drug smuggling operations, and repeated "whamming" of the airwaves with obscene language. "Some of the cases are interesting, to be sure," one FCC official commented, "but from what we've been seeing lately, CB is actually doing more for law and order than it is for the criminal." Never the less, there is certainly no shortage of hair-raising, true-life tales in the FCC's growing crime files.

SAN FRANCISCO, CALIFORNIA (from FCC files). Two CBers who had been battling for months on the air waves over who had exclusive rights to the handle, "Channel Hog," decided to fight it out in the parking lot of a large suburban shopping center. Swearing obscenities and insults at each other over the air as they raced to their meeting, the two Hogs were well primed for what turned out to be a bloody showdown.

To the horror of over 100 noon-time shoppers, both Channel Hogs drew guns and by the dual's end, the argument had become academic. Only one Channel Hog was still standing, and he was soon under arrest for murder.

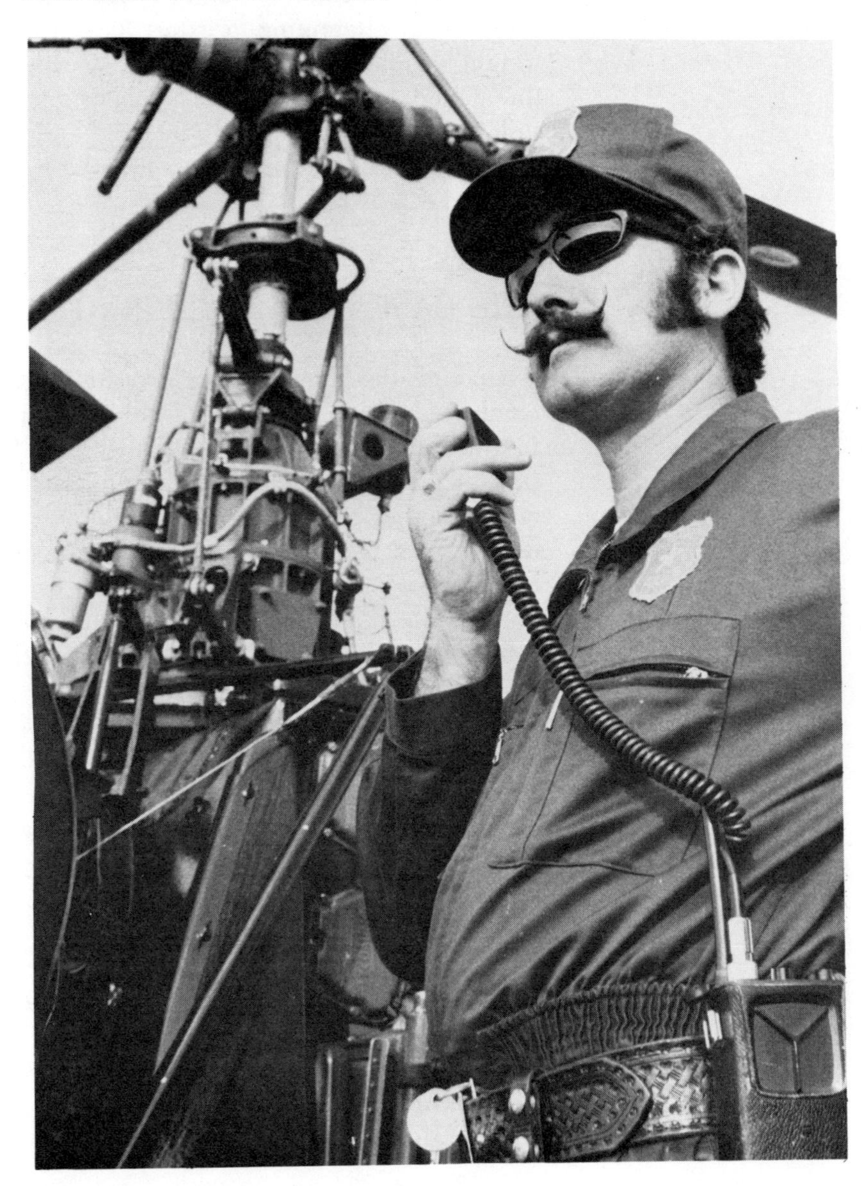

LOS ANGELES, CALIFORNIA (from FCC files). "Good Deal," a California bookie, thought he had the perfect system. With a special phone patch and a walkie talkie, he had rigged up a system so that he could make phone calls with his CB, reasoning that even if his phone was tapped and traced to his empty apartment, he would not be discovered. The deal went sour when his good CB neighbors overheard his bookmaking operation and turned him in.

NORFOLK, VIRGINIA (from FCC files). When "Bricklayer's" wife ran off with another man, it was bad enough. But when Bricklayer discovered she had taken his Golden Eagle transceiver along, it was too much. Determined to reclaim his CB, if not his wife, Bricklayer tracked her down. In the argument that resulted, he was shot and killed.

REDDING, CALIFORNIA (from FCC files). A 20-year old man was arrested for attempting to extort four million dollars from a local freight company. Communicating on his CB radio, the suspect threatened to blow up the company's main switching terminal unless his ransom demand was met. Using DF equipment, the FCC quickly got a fix on their suspect, who was quickly captured and arrested.

THE CASE OF THE BEEPING BOMBER *(from FCC files)*

One of the most bizarre extortion attempts in history was conducted with a small, hand-held CB radio in Portland, Oregon, in the fall of 1974. In September and October, eleven power transmission line towers were mysteriously dynamited east of the city. Another three towers were mined with powerful explosive devices that had failed to go off. The bombings were the work of an extortionist who identified himself as "J. Hawker," known in both the local and national press simply as "The Bomber." In his first extortion letter—addressed to the Bonneville Power Administration (BPA), but mailed to the Portland office of the FBI—he threatened to plunge the entire city of Portland into total darkness unless his demand for a one million dollar payment was met. The letter, whose contents were leaked to the press causing considerable public alarm, also threatened to set fire to Portland's Bull Run Watershed, which contained the city's water supply.

On October 20, the power company, in a public announcement, refused to pay even one cent of the million dollar demand. On October 28, the frustrated Bomber sent another letter to the FBI with explicit instructions detailing a complicated "beep" code to be used in transmitting instructions for delivery of the money over CB Emergency Channel 9.

Following this letter, the FBI notified the Portland Field Office of the FCC of the situation and requested its staff to attempt locating the station which was transmitting the extortion demands.

The Portland Office responded immediately with around-the-clock monitoring using highly sophisticated FCC DF equipment. Additional support came from a team of FCC engineers at the Seattle Field Office. Before deploying the DF vehicles, the FBI equipped them with special mobile radios with speech scramblers so their communications could not be overheard.

On October 29, the specified day for contact, the entire FBI–FCC force was positioned at strategic locations around the city. At 1 p.m., The Bomber made an extremely brief, beep-coded communication, but it was long enough for the DF-equipped vehicles to get a rough fix on his location. The FCC radioed the information to the FBI which converged in southeastern Portland, but could not locate the station which was now believed to be mobile.

On October 30, The Bomber made another brief coded contact, but the mobile signal was so weak it was impossible to DF the transmission. The FBI initiated the next contact on November 1, and this time the FCC monitors were able to pinpoint the extortionist's location at a specific Portland street intersection. The fix was transmitted to the FBI which converged on the area. One FBI car noticed a blue, badly dented Plymouth heading down the street in the opposite direction. Protruding from a rolled down car window was a CB radio antenna. Hot on the trail, FBI agents failed to apprehend the suspect when they couldn't make a U-turn quickly enough in the dense traffic. Tension mounted day by day, the public grew more nervous, and the FBI intensified its efforts, literally blanketing the transmission area with unmarked FBI cars, some brought from as far away as San Francisco.

On November 7, after six days of silence, The Bomber instructed the FBI in a letter to establish communications on Channel 9. At 1:02 p.m., according to the official FCC report, the federal agents complied with The Bomber's instructions and transmitted a coded contact phrase. The extortionist replied with four beeping dashes. A repeat second transmission by The Bomber allowed the FCC DF-vehicles to fix the exact area of operation and again the fix was immediately relayed to the FBI agents. Agent Regis Boyle, who was cruising in the vicinity of 86th Avenue and Market Street, reported monitoring the mysterious Bomber beeps, "loud and clear." FCC files show Boyle then spotted the suspicious dented, blue Plymouth and followed the car at a distance as other FBI cars rushed to the scene.

The car, driven by David J. Heesch, accompanied by his wife and one of their children, was pulled over. Inside was a walkie talkie along with a hunter's duck call (which turned out to be the source of the ominous "Bomber Beep" coded messages). Also in the car was a Luger and a loaded rifle. A copy of the transmission code previously sent to the FBI was found in the glove compartment, along with several boxes of ammunition. The Heesches were promptly taken into custody. Evidence found in the car, along with items discovered in the Heesches' suburban home, provided ample proof that David Heesch was indeed, "J. Hawker—The Bomber." The Heesches pleaded guilty and the judge sentenced Heesch to twenty years in prison and his wife to ten. The judge indicated that Mrs. Heesch, mother of three, might be released at an earlier date, but, with a stern expression on his face, said, "Mr. Heesch's sentence is not likely to be reduced by me."

UCBA President Goes To Jail *(From FCC Files)*

To hundreds of CBers, George Bennet, Founder and President of the United CBers of America (UCBA) seemed like a good egg. His club was issuing CB licenses in only one or two days, while it took months to get one from the FCC in Washington. What hundreds of innocent CBers didn't know was that Bennett and the UCBA didn't have the authorization to issue laundry tickets, much less Class-D CB licenses.

In a trial which lasted over four weeks in Detroit's Federal District Court, Bennett and the UCBA were convicted of the following: counterfeiting and distributing licenses purported to have been issued by the FCC; mail fraud; conspiracy to defraud members of the public and the United States Government; impeding and interfering with the lawful regulatory functions of the FCC; operating a radio station without a license; and filing a false application for a license with the Citizens' Radio Service.

The case was prosecuted by Gordon S. Gold, Assistant United States Attorney for the Eastern District of Michigan. Bennett was sentenced to eighteen months in prison and his club was fined $5,000. After his release, Bennett was placed on two-year probation. All damaging evidence had been collected after an exhaustive FCC investigation.

"Satan Mouth" Strikes Out *(From Combined Services)*

In New York City, where foul language is as common as the cockroach, a group of CBers decided it had no place on the airwaves, especially when their late-night chatter was being drowned out by a super-powerful base station operator who was "walking" all over them with earsplitting profanities.

"Satan Mouth," later identified as Arnold V. Press, a doubleknit textile salesman from Queens, had been "whamming" the airwaves for months with detailed stories of sexual abuse. While a few CBers reported that they found him "more amusing than late-night cable TV," many more were outraged. A CBers possee, using sophisticated DF equipment, tracked the base station to a hi-rise apartment building. They then alerted the authorities who found Press dressed in a black robe in his large studio apartment furnished only with piles of pornographic literature and thousands of dollars worth of illegal telephone and CB equipment. When taken into custody, Satan Mouth was reported to have said:

"Somebody had to tell them the truth, and the devil made me do it!"

Bust At The Bruno Bowl *(From Combined Services)*

Aaron Straus was moonlighting at the Bruno Bowl, a small-town bowling alley in Colma, California. He glanced up at the large Coors Beer clock at the end of the alley. It was 1 a.m. on a rainy Sunday morning and because the assistant manager was out sick, Aaron had manned the desk alone during the long Saturday night shift. He carefully counted out the night's proceeds: $1,500 and change. Aaron could hear water running twenty-five feet to his extreme left as Jimmy the bartender wiped up the spilled beer that covered the graying formica counter.

"Hey Jimmy," Aaron shouted, "bring me another Bud."

Aaron took the bartender's silence for assent and continued to fill out a bank deposit slip when a voice from behind him said, "Here's your beer, fella."

"Thanks," Aaron replied, half turning and extending one hand out in expectation of the cold bottle. A chilling flash of fear filled him as he realized that the voice was that of a stranger. He spun around and then stumbled back when he saw that the "beer" was a double-barreled shotgun held in a stranger's hand. It was pointed directly between his eyes.

"Easy there," the grim-faced stranger said, "one false move and I'll blow your brains out. And don't expect any help from your buddy at the bar. My friend's relieving him of his cash right now."

As Aaron headed for the register, he noticed Glenn David, a fellow CBer, coming out of the men's room at the opposite end of the alley. Desperately hoping he could signal Glenn before the robbers noticed him, Aaron raised his hands above his head with a deliberate, almost theatrical motion. "Am I supposed to raise my hands?" he asked loudly.

'Get those hands down and no more funny business!" the man snarled, pressing his shotgun into Aaron's stomach. But the trick worked and Glenn returned to the men's room where a small window lead out into the parking lot. Glenn's CB-equipped car was parked outside.

In the dark parking lot illuminated only by the thin crescent of a waning moon, Glenn quickly turned on his CB radio.

"I have a 10-33," he shouted. "Emergency at the Bruno Bowl, come back."

On emergency monitoring duty at his home base, Bill Lee answered immediately.

"What's your problem, come-awn."

"Armed robbery in progress at Bruno Bowl. Alert Colma Police and California Highway Patrol."

"10-4, will call immediately. And be careful."

Meanwhile back at the Bruno Bowl, Aaron had handed over the money from the register and the two robbers were backing out of the alley. Crouched in his pale green Volkswagen which he had pulled out in front of the Bruno Bowl, Glenn watched the two men escape in a blue Vega.

"Bill," he called out to the base station. "They're escaping and I'm going to follow them."

"Stay back," the base station warned, "but keep me informed of their 20. I've got the police on the land line."

Following close behind the Vega, Bill kept in constant contact with the base station. Within minutes a police car was speeding up from behind and within seconds, the Vega had been pinned against the edge of the road.

Later the two men were taken back to the Bruno Bowl where a shaken Aaron Straus identified them. The owner of the Bruno, who had been summoned to the scene, commended Aaron and Glenn for their quick thinking and brave action.

FCC officials in Washington are usually uneasy about publicizing sensational cases from their crime file, because the total CB picture they say emphasizes more assistance from CB to the forces of law and order than to crime. "Take the simple case of a highway patrol car," argues FCC agent Richard Smith, Chief of Investigations, "CB has expanded the effective range of that car by a factor of twenty. It used to be that a speeder slowed down when he *saw* a patrol car and then speed up again a few miles later. Now, because of CB, they *hear* about the police way in advance and slow down for a much longer stretch. Add to that the thousands of CBers who actually work with law enforcement officials and you have a pretty good idea just how helpful CB really has become."

The attitude of law enforcement officials has changed dramatically since the early 1970s when antagonism between police and truckers with CBs was at its height. Now more and more police chiefs are arguing the CB case: those additional "eyes and ears" out on the highway and in the communities are for the good of everyone.

WHEN **CBERS** LEARNED THAT TWO **CONVICTS** HAD **ESCAPED** FROM AN IOWA **JAIL** THEY TOOK TO THE **AIRWAVES** AND ORGANIZED A

MANHUNT!

John Butterfield 3/76

A CBER WITH A RIG IN HIS PICK-UP HEARD THE DESCRIPTION!

THERE THEY ARE!

I'VE GOT TO HEAD THEM OFF!

SKREEE

MINUTES LATER THE POLICE ARRIVED AND CAPTURED THE ESCAPED CONS--THANKS TO A BRAVE CBER!

dick Siegel '76
RIPPED FROM THE CRIME FILES OF THE F.C.C!!
THE HEIST
DATE: AUG. 22, 1975
PLACE: TEXAS NATIONAL BANK
TIME: 2:45 P.M.
NK

BANK

LET'S GO!

EVERYBODY FREEZE!

DROP THAT GUN OR I'LL SPRAY HER BRAINS ALL OVER THE BANK!!!

HURRY UP MR. TELLER!

WHAT THE...?
CB BIBLE

MAYDAY 10-33 EMERGENCY! THE BANK'S BEING ROBBED!
2 MEN WITH GUNS! THEY HAVE LISA HOSTAGE, MAYOR!
ON MY WAY, JIM. I'LL PICK YOU UP!
LISA MY DAUGHTER! THE FIENDS!

LET'S GO!

GOOD!
THEY'RE HEADING NORTH! I CALLED THE POLICE!

WE'RE BEING FOLLOWED! PUT THE PEDAL TO THE METAL!
EAT LEAD, YOU TURKEYS!!!
BLAM!

UH-OH...

THE POLICE! OH GOODY!
SHUT UP, YOU FOOL!

SURRENDER AND NO ONE WILL GET HURT!

WE GIVE UP!

DAD!

HEY CHIEF! LOOKEE HERE!
FOILED!
U.S.
WELL, IF IT ISN'T FRED SMITH THE DEPUTY FROM OVER IN WACO COUNTY!

DON'T THANK ME! THANK MY CB!
NICE GOING JIM!
The End

CB to the Rescue: Rise of the Emergency Assistance Organizations

Every society has its criminals, and as we've seen CBers are no exception. But most law officials now agree that the advantages of CB far outweigh the disadvantages. Even in the case of truckers who still use radios to avoid speed traps, they are, in other respects, not only law abiding, but law assisting. Colonel S. S. Smith, Superintendant of the Missouri State Highway Patrol and a pioneer in supporting CBers, claims that "truckers are proving to be a valuable ally in ridding the highways of intoxicated drivers and providing assistance to motorists in trouble. Since we equipped our patrol cars with CB radios, we estimate that more than 20 percent of all accident reports now come in on the CB. And CBers are helping out in emergencies of all kinds in increasing numbers."

When CBers learned, for example, that two convicts had escaped from the Mills County (Iowa) jail, they took to the airwaves and organized a manhunt with the blessing of the Freemont County Sheriff's office. In fact, the police asked the CBers to pass on a description of the getaway car to 18-wheelers on Interstate 29. When a trucker spotted the car on the interstate, he relayed the message to the local Smokies and the chase was on. The escapees were eventually trapped by a brave CBer who sacrificed his pickup for a roadblock and dove for cover in a roadside ditch. Minutes later, the police arrived and captured the blocked convicts.

As early as 1964, civic-minded CBers (many of them members of a young organization called REACT (Radio Emergency Affiliated Citizens Teams) had been using Channel 9 as an emergency channel. But it wasn't until the CBers organized and petitioned the FCC that the channel was legally limited to "emergency communications involving the immediate safety of individuals or the immediate protection of property or communications necessary to render assistance to a motorist." This became effective on July 24, 1970, and the prior voluntary use of the channel for emergency purposes had strong bearing on the FCC's decision. A REACT study showed that as of 1966, 1,800,000 incidents had been handled on Channel 9, including close to 500,000 automobile accidents. In effect what the FCC was doing was recognizing a *de facto* emergency channel that had developed through the wholly voluntary efforts of thousands of CBers.

At about the same time Channel 9 became official, REACT officials helped initiate a program with the Ohio State Highway Patrol, equipping both police cars and police stations with CB transceivers. In a pilot study, it was found that cooperation between CBers and police had significantly reduced the time needed to respond to an accident. The Ohio experiment and a similar one in Missouri were so successful that a host of other states are either using or experimenting with CB equipment in their highway police cars and stations. To date the list includes California, Georgia, Idaho, Indiana, Maryland, Minnesota, Mississippi, New Jersey, New Mexico, Oklahoma, Oregon, Pennsyl-

FOR RELEASE AT WILL

Indiana State Police Superintendent Robert L. DeBard announces that the Department has now been issued a license by the FCC for operation on citizens band frequencies.

Most Indiana State Police vehicles are equipped with C.B. radios, purchased at the troopers' expense. All troopers are authorized to operate under the call letter of KFP 2179.

Since troopers normally monitor Citizen Band Channel No. 10, any motorist or operator of a C.B. unit wishing to report a hazardous or emergency situation can call for an Indiana trooper and request he tune to C.B. channel No. 9.

Channel 9 is specifically reserved for emergency communications involving the immediate safety of life and protection of property, or to assist motorists in need of help.

"The State Police commend the truckers with C.B. radios in their rigs for their response to fellow motorists in trouble on the highways. They have helped in countless situations by radioing for emergency vehicles, and many lives have been saved because of our ability to respond to critical situations with great speed," stated Superintendent DeBard.

To expedite matters when attempting to contact a trooper, every effort should be made to use the assigned call numbers of KFP 2179.

WANTED

The Indiana State Police are searching for two truck drivers to personally thank them for their vital help in apprehending the killer of a Crawfordsville (Indiana) police officer.

The victim, Lieutenant Russell Baldwin, 44, was shot and killed by an armed bandit following a holdup in Crawfordsville on the night of August 27, 1974.

A police manhunt forced the killer to hide in a wooded area until late the following afternoon when the suspect attempted to hitch a ride on Interstate 74. He was spotted by a passing trucker who used his citizen's band radio to call a nearby base station with a description and location of the suspect.

An unsuspecting motorist then stopped to give the killer a lift. Another trucker witnessed the incident and radioed the base station a description of the car.

The base station operator, Nina Curtis, 201 West Chestnut Street, Crawfordsville, telephoned both reports to the State Police. Within minutes, troopers pulled the car off the road and the suspect surrendered.

If you're one of the truckers the Indiana State Police are searching for, we urge you to "surrender." The troopers would like to shake your hand, probably present you with a certificate of appreciation and, maybe, even buy you a cup of coffee.

vania, Rhode Island, Tennessee, Utah, Washington, and Wyoming. In Illinois, the Independent Truckers Association has raised money and donated it to the state police to be used to equip their patrol cars with CB.

To help traveling motorists to contact police in different states, the FCC is considering a policy of identifying police CB stations with a standard "formula" of call letters, consisting of the prefix K, followed by the two-letter state abbreviation, and the numbers 0911, 911 being the emergency telephone number in most communities.

If the program policy is put into affect, every CBer could reach the Smokies with a minimum of effort, by calling KRI-0911 in Rhode Island, KTX-0911 in Texas, KID-0911 in Idaho, and so on.

The Electronic Industries Association, representing the nation's manufacturers, recently found that more than 1,000 local police stations now monitor Channel 9. A *Transportation Research Record* report analyzing the Ohio experiment concluded that "the use of CB radio for emergency communications is a substantial existing and constantly growing resource that makes no demands on public monies."

"We consider CB radios the most revolutionary idea to occur in law enforcement in this century," comments Colonel Smith, "because it not only has the potential of providing upgraded service to the police, but it will provide the means whereby the average citizen can become involved in law enforcement."

And involved is exactly what CBers are becoming, both as individuals and through association with more than 2,500 local and nationally organized emergency teams. No doubt part of the attraction is the appeal of playing "cops and robbers," and sometimes these amateur Smokies take themselves too seriously. They get so carried away that they don't help either the police or the public, and actually can hinder the authorities and give CB a bad name.

It's easy for an overimaginative CBer to graduate from being a monitor, who calls the police, to actually answering emergency calls himself. Some CBers in Southern California have painted their cars to look like real police vehicles, wear blue uniforms and habitually monitor the police frequencies. These "kiddie cops" don't impress anybody but themselves.

In Bristol, Virginia, a trucker witnessed a hit-and-run incident, picked up his CB mike, and put out a call to his 18-wheeling friends. Within minutes, a small convoy of giant semis had caught up with the suspect and were using their mammoth rigs to box in the 4-wheeler. The trapped driver was kept in a virtual mobile prison until the law arrived and arrested him. And when another hit-and-runner struck and killed a New York State trooper, a trucker radioed ahead a description of the car. The CB alert helped police apprehend a suspect.

In Twin Falls, Idaho, three petty thieves robbed a gas station ripping out the telephone wires as they escaped. But they overlooked the station's CB and the attendant used it to call for

help after the robbers left. The thieves were soon captured by state patrolmen who had been notified by CBers monitoring the emergency channel.

A less dramatic, but poignant CB assist occurred in Columbus, Ohio, when an injured welfare mother's pantry shelf was looted. A neighbor, aware of the situation, used her CB to transmit an appeal for help. Within hours, local CBers had refilled the pantry with food. In Missoula, Montana, ghosts and goblins were kept in check by a 32-person anti-vandal CB patrol which worked closely with the local police each Holloween night. The Community Radio Watch team is part of a Civil Defense program which operates on Halloween night between 7 p.m. and the early hours of the morning.

COMBATING FEAR IN THE STREETS

Fear in the streets is epidemic in many American cities, but now angry citizens are learning to fight back with the help of portable CB gear. In the Asylum Hill neighborhood of Hartford, Connecticut, elderly citizens who had been victimized by purse snatchers, robbers, and street gangs organized their own "Street Watchers" program. Every day a dedicated core of silver-haired ladies and gentlemen, armed with walkie talkies, watched over the neighborhood. Sitting in front of their homes or manning window posts, a call went out to the base station at the first sign of trouble. Since implementation, street crime in Asylum Hill is down by 80% and residents report that for the first time in years they can walk their streets and feel secure. City officials admit that the CB program is more effective than increasing the number of police patrols in the area.

On New York City's Upper West Side, where shootouts between police and criminals happen all too often, the Citywide REACT team has taken on a dangerous mission. Their walkie-talkie equipped foot patrols scout this dangerous part of town, covering up to fifteen square miles a night. Working with tenant patrols, neighborhood guards, block associations and police, the Citywiders have alerted police to dope dealers and pistol-packing criminals. Dr. William Wahlin, President of Citywide and a psychology professor at Bronx Community College, started the organization after the Harlem riots in 1968. He's convinced that the team's efforts have both helped reduce crime and broken down some of the community alienation.

Urban CB activity isn't always of the cops and robbers variety. When a Miami boy was born with serious facial deformities, local CBers organized a fund drive on the airwaves. They were able to raise close to $10,000 for the expensive plastic surgery. When earthquakes left thousands homeless in Guatamala in 1976, CB clubs across the country organized fund drives and sent needed supplies to the disaster victims.

National Emergency Organizations

During earthquakes, floods, tornados and hurricanes, when telephone lines are down and disaster victims are in need of immediate assistance, emergency communications become essential. Until recently it was the ham who carted in his portable equipment and relayed messages in and out of a stricken area. But today, CB emergency teams are on the scene moments after catastrophe strikes. Highly mobile, often well trained and organized CBers are able to report changing conditions and maintain contact with rescuers working in the area. There are hundreds of independent emergency groups in the country, but many choose to affiliate themselves with one of the large national organizations like REACT, ALERT, or REST-MARINE.

Founded in 1962, REACT is the oldest, largest and most influential of CB organizations. An independent, nonprofit corporation, REACT is still supported in part by General Motors, one of its oldest sponsors. Headquartered in Chicago, REACT Managing Director Gerald H. Reese helps coordinate the activities of more than 1,000 teams in all fifty states and seven Canadian provinces. REACT teams have assisted at every major natural disaster in recent U.S. history, including Hurricane Camille, and the Rapid City, South Dakota flood where five REACTers lost their lives performing emergency service. In the short history of the organization, over 200,000 volunteers have provided a total of more than 100 million man-hours of free public service through participation with local REACT teams. REACTers have handled an estimated 60 million emergency calls, including over 13 million highway accidents.

When not on emergency duty, a typical REACT team like one in Ft. Lauderdale, Florida, will perform more than a dozen motorist assists and serve close to 2,000 cups of coffee to fatigued

REACT

REACT, a nonprofit organization, is the oldest and largest affiliation of radio emergency teams. These are the official REACT goals.

(1) To develop the use of the Citizens' Radio Service as an additional source of communications for emergencies, disasters, and as an emergency aid to individuals;
(2) To establish 24-hour volunteer monitoring of emergency calls, particularly over officially designated emergency channels, from Citizens' Radio Service licensees, and reporting such calls to appropriate emergency authorities;
(3) To promote highway safety by developing programs for providing information and communications assistance to motorists;
(4) To coordinate efforts with and provide communications help to other groups eg, Red Cross, Civil Defense and local public authorities, in emergencies and disasters;
(5) To develop and administer public information projects demonstrating and publicizing the potential benefits and the proper use of Citizens' Radio Service to individuals, organizations, industry and government; and
(6) To further the above purposes by chartering local Radio Emergency Associated Citizens' Teams which will carry out programs implementing the purposes of the corporation on a local basis.

A brochure about REACT is available on request from REACT Headquarters, 111 East Wacker, Chicago, IL 60601.

---

motorists on a three-day holiday weekend. REACTers also help police control parade traffic and crowds. Often they organize fund drives for local community projects.

But REACT's most important full-time activity, and a contractual agreement between the local team and the national organization, is to make every effort to monitor Channel 9, 24 hours a day, seven days a week.

REACTers enjoy their work and take it very seriously. In Norfolk, Virginia, the Tidewater REACT Team has a comprehensive program complete with Air and Marine Wings. Their arsenal of CB mobile units includes a privately owned airplane, three boats, six motorcycles, three trucks and a team-owned specially-equipped van with emergency and rescue equipment. Many REACT teams, like the Redicom unit in Illinois have had both Red Cross and Civil Defense training and are qualified to provide emergency medical and rescue as well as communication service. Today, REACT has a working agreement with the American Red Cross and hundreds more REACTers are receiving medical training. The Civil Defense Preparedness Agency in Washington, D.C., is now providing matching funds for CB base stations that qualify to assist in Civil Defense activities.

In cities, REACT teams often work to combat crime and make

emergency assists. Pierre Furst, a soft spoken United Nations employee by day, becomes a dedicated REACTer every night. Patrolling the streets in a station wagon he converted at his own expense to an ambulance, Mr. Furst constantly monitors Channel 9. In 1975, he pulled eight people out of serious wrecks and delivered a baby. "As long as I can save one life," he said, "it's worth the work."

Smaller than REACT, but equally dedicated and active is ALERT (Affiliated League of Emergency Radio Teams). ALERT's activities are basically identical with those of REACT — emergency assistance, motorist aid, and community service. But Robert Thompson, President of ALERT, also sees his Washington, D.C. headquartered organization as something of a CBers lobby. On behalf of the 14,000 ALERT members and 600 ALERT teams, Mr. Thompson is taking the FCC to court. Citing a Supreme Court decision, for example, rebating license fees in excess of actual processing costs to cable TV station operators, Mr. Thompson contends "CBers have the same rights as anyone else dealing with the FCC for licensing." He's seeking a $16 rebate to all the CBers who paid $20 for their license before the price was lowered to $4 to reflect actual processing costs.

While ALERT is basically similar to REACT, there are two basic differences: REACT is a nonprofit corporation; ALERT is not. REACT only accepts team memberships (five or more persons), while ALERT solicits individual participants.

Founded in 1971, REST-MARINE (Radio Emergency Service Teams) specializes in marine service. With only 75 members and 4 teams, REST-MARINE is considerably smaller than the other national organizations, and in many instances, its services are duplicated by the Coast Guard which uses VHF/FM equipment. But REST has a legitimate concern. Coast Guard equipment is fine where available, but there are many lakes and rivers not within Coast Guard jurisdiction. In these areas, REST-MARINE sees a real need for a marine CB emergency channel to help assist lake and river craft in trouble. REST-MARINE teams have already proven themselves effective in the Potomac River around Washington, D.C. and on Lake Erie, near Cleveland and Toledo.

There are dozens of effective, independent emergency organizations throughout the country, some with unique programs. In Douglas and Sarpy Countries, Nebraska, one club maintains an effective tornado warning system. In Pennsylvania, HAM (Highway Assist Modulators) patrols the turnpikes. The services of all the independent organizations are too numerous to mention, but they do prove that CB is neither a fad nor a hobbyist's electronic toy. The national and local organizations have established beyond any doubt that CB creates an opportunity for people to work together, strengthen community spirit and provide valuable public assistance. Colonel Robert M. Chiaramonte, Superintendant of the Ohio State Highway Patrol, recently concluded, "The day is coming when it may be said that help for someone in trouble is truely as close as his microphone."

Big Juicers

"CB is bringing back the good old lively art of conversation and people are becoming more articulate in a respect they might never have before."

Bob Cole, the "Midnight Rider"
WWOK, Miami

"Break, break, break, break."
"Come on, breaker."
"You got the one Buddy Ray here. Who's that out there, come on."
"You got Sweet Blue Eyes."
"Sweet Blue Eyes?"
"Ten-fer."
"I don't believe I ever talked with y'all. What's your 20, Blue Eyes?"
"Right behind Martinsdale High School.'
"You mobile?"
"No, I'm a base."
"Watcha doin' up at 3 a.m. Blue Eyes?"
"Just home from work."
"Where y'all workin'?"
"Ponderosa"
"Ponderosa Steak House?"
"Ten-fer."
"You gonna cook me a good piece of sirloin, y'all hear:
I get over there sometimes. Listen, Blue Eyes,
I'll catch y'all later. I gotta go play radio."

With 50,000 watts of clear channel power crackling behind him, Big Juicer Buddy Ray of WWVA pushes his country music program through the night, driving those white man's blues into the crisscross of northeastern interstates, way up into Canada, down south as far as Charlotte, and east into the Big Apple, from midnight till dawn, six days a week. But when it comes to personal pleasure, those 50,000 watts can't hold a candle to the 5 watts pumped out by Buddy's CB, a top-of-the-line transceiver located right in the studio, not ten feet from the illuminated dials and gadgets on his large AM broadcast console. A maximum-legal height CB antenna sits atop the Capitol Theater in Wheeling, West Virginia, home of WWVA and the weekly jamboree Buddy emcees. "'Course I have to be real careful in here," Buddy explains with a boyish grin, "there's no way in the world WWVA or the FCC will allow one discernible word from that CB set to get on the station's airwaves. But that doesn't mean I can't modulate with my friends between songs."

The Kentucky-born disc jockey with a Mick Jagger smile and clumsy, irresistible boyish charm, knows how to gear a radio program to the truckin' audience with the precision of a driver shifting all the way from granny to going-home gear. A country boy who grew up with relatives in the "truckin' business," Buddy is as at home

WWVA's Buddy Ray.

with his audience as a spark plug in an engine. He talks their language, knows their music, and even more important, modulates with them on his CB. On a typical night with a Dave Dudley song on the AM deck, Buddy will swing around, abandon his broadcast chair, and grab hold of his CB mike.

There are a lot of calls for Buddy Ray on his CB and its not unusual for a local CBer, sometimes sweet young things like Red Streak, 'Lil Rabbit, or Bull Doggie, to come into the studio in the wee hours of the morning for a visit with their CB buddy and favorite night-time DJ. They'll come alone, or in groups and bring Buddy a late-night sandwich and coffee snack. "The CBers around here are real friendly folk," says Buddy, "and that's one big reason I enjoy CB so much." One time, as Buddy's antenna caught the first rosey-colored fingers of dawn, a mammoth 18-wheeler parked outside the Capitol and the gear jammer came in to eyeball the man he'd been listening to and modulating with on his CB. "I love it. There's a whole crazy world out there on the CB and sometimes it walks right into my studio."

When Buddy moved to Wheeling, in 1971, to do the all night truckers' show, "it only seemed right that I should bring my CB along." There was some argument with the station about having it in the studio, but Buddy won out. "It's the best way for me to talk directly with the people out there listening, especially with the truckers on the road who can't stop to make a phone call, but want to request a song."

Spinning records and modulating, it's all part of what Buddy Ray calls playing radio and he's been playing with some form of radio all his life. "Me and CB go back a long, long time," Buddy recalls. "When I was a kid, a friend of my daddy's, Mr. Wilson, had himself a ham set and that's when I first wanted to become an engineer. I started out studying electronics, got myself a degree, then worked awhile repairing CB units and finally I became a DJ."

Though Buddy's bashful about being the first all-night Big Juicer to bring a CB to the studio, he's partly responsible for the growing popularity of CB. His double-threat radio role has inspired a host of other juicers around the country to follow his example. "Now, I didn't really have that much to do with the popularity of CB," Buddy insists, "it's the truckers' thing and that's where the credit belongs. I'm just along for the fun of it. CB's the largest bill-free telephone in the world."

But when Buddy built a studio shelf for his CB, the idea went right into the stratosphere, and a host of other DJs picked up on it.

Down in the Bikini State, you'll hear the "Midnight Rider," Bob Cole of WWOK, Miami. One of the bright young stars on the airwaves, Bob is a fast-talking, on-target juicer who's become the favorite with the new breed, younger trucker. Bob's articulate style and soaring success is the envy of many a juicer out to catch the gear jammers' attention: "You got the 'Midnight Rider' here, the all night, hill billy, disc jerkey, kiss stealin', wheeling

WWOK's Bob Cole.

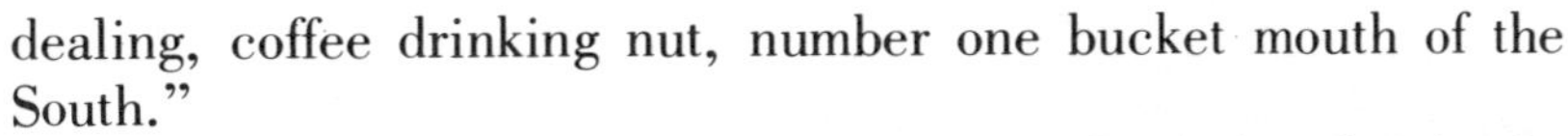

dealing, coffee drinking nut, number one bucket mouth of the South."

A college student cracking books by day, the "Midnight Rider" sounds ten-years older than he is, yet he's as close to the trucking spirit and as dedicated to truckers' music as any man in the business. With a family tree that blooms relatives from Knoxville to Nashville, from Macon to Louisville, Bob's country credentials are "clean and green" all the way down the line. Ever since he was seven years old and went to visit his DJ cousin in Oak Ridge, Tennessee, Bob knew his career would be in radio.

The "Midnight Rider" discovered CB more than four years ago and has had one in his truck ever since. In 1975, he brought the CB into the studio and like Buddy Ray, he welcomes song requests and road condition reports over his CB and receives plenty. "I'm basically a redneck, man. I drive a pick-up, get off talking to truckers, spending my time at truck stops, ratchet jawing on the CB. That's where my head's at."

Bob appreciates the safety and public service value of CB, but in his opinion it's made an even more significant contribution. "CB is bringing back the good, old lively art of conversation and people are becoming more articulate in a respect they might never have before. Without having to eyeball you, shy people are learning how to talk and carry on a conversation. Old ladies who used to sit alone, holed up in their houses, are coming out on the airwaves and ratchet jawing with the rest of us. CB is more than a fad. It's just beautiful."

About 700 miles northwest of WWOK, is WWL, New Orleans, home of Charlie Douglas, granddaddy of the all-night trucking programs (which he originated in the late 1960s). One of the all-time trucker favorites, Charlie would have a CB in the studio, but with more than 200 phone requests jamming the lines each night, he doesn't know how he could handle the additional calls.

Big Juicer John Trimble ("Real Niterider") has a good reason for not having a studio CB, though he has one in his car. His station, KWKH in Shreveport, Louisiana, beams the show right out of a truck stop, "where there's so much chatter on the CB no one could get a call to me." When a trucker wants to request a song, "he either picks up the telephone or walks right into the studio."

WBAP s Bill Mack

Heading out of Shreveport, along Interstate 20, you know you're approaching Dallas when the gas stations become as thick as license plates on a K-Whopper's backside. Like his fellow juicers, Bill Mack ("Midnight Cowboy") on WBAP sends out trucking songs till the sun comes over the horizon. But the "Midnight Cowboy" has the same problem as many CBers in the larger urban areas: so much chatter and noise on all the channels that it's practically impossible to carry on a conversation. But in Amarillo, Texas, where there's a lot less folk per square mile and a lot less chatter on the airwaves, "Country DeeJay," West Whittle, on KDJW uses his studio CB as only a Big Juicer can: fielding song requests and relaying road and emergency reports to the gear jammers all night long.

KVOO's Billy Parker.

If you tune in your CB in Tulsa, you may catch Bill Parker ("Average Man"). Winner of the Country Music Acadamy Disc Jockey of the Year award in 1976. A singer who toured with Ernest Tubb and still makes recordings, Average Man (who has a country song out with the same name) has three CBs, one in his car, one in his van, and one in his home. And Average Man West's thinking about getting one for his station, KVOO.

Following Interstate 40 due east out of Tulsa, you'll eventually pick up the Hairl Hensley Show on WSM in Nashville, Tennessee. And if you keep on moving east, it won't be long before you catch Larry James, complete with CB, on WBT in Charlotte, North Carolina.

Wichita, Kansas, is in the center of America and there, where the airwaves are as clean as the chrome on a cowboy's rig, you'll find another great truckin' DJ with a CB in the studio. Pig raiser and wheat farmer by day, Buddy Nichols ("Mulvane Mouth") has been broadcasting to the truckers for close to 11 years and has had that CB in the studio almost as long as Buddy Ray himself. Unfortunately, his station, KFDL only puts out 1,000 watts nighttime, and though this cuts down on range, it also reduces the phone request traffic and give Nichols more time to enjoy modulating on the CB.

KLAC's Chuck Sullivan.

Heading west out of Wichita, it won't be long before you can pick up the powerful signal of KLAC stretching out from Los Angeles where Chuck Sullivan hosts his own special truckers' program. "Our station ran a handle contest for me," Sulliv n said, "and I looked at hundreds of them before I chose 'Music Maker." Sullivan only recently put his CB transceiver in the studio, but if you can penetrate the Los Angeles chatter, he's ready to modulate.

Far from the smog and glitter of Los Angeles, up north in God's Country, you can pick up Steve McNally ("Ozark Oakie") on KWJJ out of Portland, Oregon. "If you are mobiling and have road condition reports or other information, give me a shout on your CB." Further north, you can hear Patty Parr on KGA in Spokane, Washington. The only female truckers' DJ on the air, Patty's a college student who "more or less stumbled" into the radio game. Patty loves truckers and country music and "One day," she said, "the station manager suggested I use some CB lingo on the program. I love the sound of it and I went out and got myself a CB and began to practice."

WMAQ's Fred Sanders.

Heading back east along the northern route, you'll pick up the entertaining Billy Cole Show coming out of Des Moines, Iowa. Then before long, you'll hear the clear 50,000 watt signal of WMAQ from Chicago, Illinois.

With a multimillion dollar corporation behind him, popular "Big Juicer" Fred Sanders is the highest paid and probably best informed truckers' DJ on the air. He gets weekly reports on the most requested songs and items of interest to his audience. Fred, who has had a CB in his Porsche for over a year, picked up the handle "Fast Freddy" while putting the pedal to the metal and

Where to Find the Big Juicers

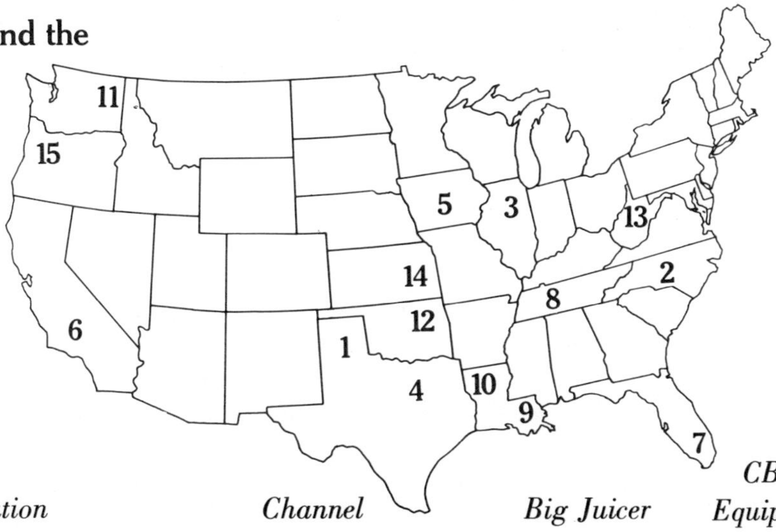

	Location	*Channel*		*Big Juicer*	*CB-Equipped*
1.	Amarillo, Texas	KDJW	1010	West Whittle "Country D.J."	yes
2.	Charlotte, North Carolina	WBT	1110	Larry James	yes
3.	Chicago, Illinois	WMAQ	670	Fred Sanders "Fast Freddy"	yes
4.	Dallas, Texas	WBAP	820	Bill Mack "Midnight Cowboy"	yes
5.	Des Moines, Iowa	WHO	1040	Billy Cole	no
6.	Los Angeles, California	KLAC	570	Chuck Sullivan "Music Maker"	yes
7.	Miami, Florida	WWOK	1260	Bob Cole "Midnight Rider"	yes
8.	Nashville, Tennessee	WSM	650	Hairl Hensley	no
9.	New Orleans, Louisiana	WWL	870	Charlie Douglas	no
10.	Shreveport, Louisiana	KWKH	1130	John Trimble "The Real Niterider"	no
11.	Spokane, Washington	KGA	1510	Patty Parr	yes
12.	Tulsa, Oklahoma	KVOO	1170	Billy Parker "Average Man"	no
13.	Wheeling, West Virginia	WWVA	1170	Buddy Ray	yes
14.	Wichita, Kansas	KFDI	1070	Buddy Nichols "Mulvane Mouth"	yes
15.	Portland, Oregon	KWJJ	1080	Steve McNally "T.C."	yes

cruising along at 100 mph in a convoy. WMAQ gave the green light for an in-studio CB for Fast Freddy in March of 1976 and Freddy's been modulating ever since.

Truckers' shows don't exist in New York City, but country fans have found their music on WHN, the first AM outfit to call itself "Your CB radio station." WHN has been doing a good job spreading the gospel by featuring CB songs and mailing free CB lingo cards to their listeners. In January, 1976, they received 50,000 CB card requests.

The Big Juicers connect CBers like super slabs connect cities and they keep the airwaves crackling and the truckers alert almost as effectively as 100-mile coffee. But there wouldn't be juicers if there weren't country music. And it's the music, and a few songs in particular, that are probably the single most important factor in the growing interest in CB language and culture.

Trucking Music Discovers CB

Next to the roar of a well-tuned engine and the whistle of spinning wheels along a clear super slab, there's nothing that sounds better to a trucker's ear than the twang of good country music and especially the sound of trucking songs which nowadays often are based on or incorporate a good deal of CB language.

Country music and truckers got together in 1939, when Cliff Bruner and His Boys recorded "Truck Driver's Blues," a song that romanced the ache of driving along an endless line of lonely highway. During World War II, when trucks began to compete with the railroads as a basic mover of supplies and goods, songs about Casey Jones began to be replaced by tunes about the trucker, songs like "Alcohol and #2 Diesel" and "Gooseball Brown." Trucking songs, though, never had a wide appeal even among country music fans until 1963, when Dave Dudley's hit, "Six Days on The Road," climbed to the top of the country charts.

Many country fans and most CBers don't realize that songs romancing CBs and Smokey Bears came out quite a few years before the commerical success of C.W. McCall's "Convoy" and Cledus Maggard's "The White Knight." Those early CB tunes, like Paul Click's "Smokey, Trucks, & CB," and Kenny Price's "Let's Truck Together," were a lot simpler than the CB hits of today. Written back in the good old days when most of the original 23 channels were as clean as a freshly pressed LP, truckers were about the only people who could understand them.

"Nobody knew what I was talking about when I came out with 'Let's Truck Together'," confessed Kenny Price. "My wife's brothers are all in the trucking business and they all used CBs. So my wife and boys got me a CB for my birthday. Well, I was sitting shotgun in the bus while my son was driving when I heard some truckers talking on the CB. They were coming out with all these great terms, so I grabbed my note pad and before you know it, I had something going." Donna Price, Kenny's wife, worked as much on the song as he did and the whole family participated in

Cledus Maggard

Kenny Price

the recording. Price's RCA producer didn't understand the song but had enough faith in the singer to give the go-ahead and press it. "Let's Truck Together" went on to become a success in many country markets.

More recent CB singles (many have still played only on country stations) include Dave Dudley's "Me & Ole CB," Jim Moore's "Southbound," Roger Rainy's "Breaker, Breaker," and Sherri Pond's "Calling Rubber Duck."

With siren sounds courtesy of the Tennessee Highway Patrol and truck noises courtesy of the Peterbilt Motor Company, "Talkin' Smokey" is the first authentic all-CB album. Produced by a Texas truck-stop operator and truck owner-operator for 14 years, the album was released in the spring of 1975, and is sold at truck stops and through mail order.

Composer-singer-arranger Jodie Lyons ("Music Man") spent months in the cab of Skipper Whitson's rig, trucking and modulating before he wrote a word. "To write about something in depth," he says, "requires first-hand knowledge . . . you have to be out on that old boulevard and see exactly what's going on . . . talk to hundreds of truckers . . . participate in the modulation." Recorded in Nashville on 16 tracks with 24 country musicians, "Talkin' Smokey" is as professional and satisfying as anything that's come out of Music City. And it has an authenticity that makes it genuinely appealing to the gear jammers for whom it was written.

When it comes to tapping the psychic nervous system of America, sophistication often has it over country. So it's not as curious as it might first appear that the two giant CB hits, "Convoy" and "The White Knight" weren't written either by truckers or country and western artists, but by ad men, people whose systems are tuned to the tempo of the times. Not that either of these songs are pure commerical contrivances, they're actually light-hearted, inspired experiments with a new language that surprised their authors when they overwhelmed the public's fancy.

Bill Fries (C.W. McCall), the lyricist and singer of "Convoy," has been heading for musical success ever since his catchy Old Home Bread commericals captured a coveted Clio (Best National TV Ad Award) in 1975. In the commerical (later modified and released as a the single record), a trucker named C.W. would deliver a batch of Old Home Bread to the Old Home Fill-'er -Up-and-Keep-On-Trucking Cafe where Mavis, a waitress with a soft spot for the trucker, would butter his bread. When "Old Home" was released as the single in the Spring of 1974, it went on to sell 100,000 copies. "Old Home" was followed by a second single, "Wolf Creek Pass," a humorous ballad about a truck that loses its breaks coming through the Rockies. (A third single, "Classified," a comical monologue about buying a used pickup, never really got rolling.)

In 1975, Fries wrote "Convoy," the story of 1,000 trucks speeding across the country, busting through road blocks and crashing toll gates as Smokies on wheels and in the air helplessly watched. "I wrote 'Convoy'," Fries recalls, "after I found myself in the middle of a little convoy coming out of Denver. The jargon struck me and then, of course, my imagination runs bigger than life and suddenly I pictured this convoy getting bigger and bigger till no one could stop it." Fries spent three weeks writing the song and another two recording it over again.

When Fries pushed to have "Convoy" released as a single, MGM resisted, and originally it was just another tune on his new "*Black Bear Road*" album, until a few DJs started featuring it. "MGM," recalls Fries, "then had to put four record pressing plants on the line 24 hours a day to meet the demands. The funny thing about this song is that it's a gold record in Canada, a platinum in the U.S., number one in South Africa, number one in Sweden and coming up fast on the charts in London, West Germany, and Japan. And they don't understand a word of it."

Trying to explain the popularity of his creation, the songwriter suggested that "maybe it's the rebel instinct up against the establishment. Plus the fact that it has a very militant sound and gets people's blood pressure up. And I think the world wants to be talking CB jargon. There's a magic about '10-4,' a kind of romantic mystique about the whole thing."

"Convoy" is much more than a song about CB: it's an iconoclastic call to join a counter culture, to participate in the overwhelming thrill of driving through the night in a party of such power that nothing and nobody can stop you ("Come on an' join our convoy; ain't nothing gonna git in your way"). The song doesn't really promote lawlessness, as some critics have suggested, as much as it graphically presents a breaking through of social restrictions (tolls, speed limits) in the process of pursuing one's destination. Combined with a hard driving, martial sound and a slick, strong production, the song is irresistible, even when the lyric isn't understood.

While "Convoy's" been a runaway success with both country and rock music fans, it's rattled some authorities. Governor Ray of Iowa, condemmed the song for encouraging speeding and reckless driving and a big 50,000-watt station, WHO in Des Moines, banned it for awhile. "But the thing backfired," Fries mused. "They'd been selling 200, maybe 300 records a week and after the governor came out and said he was having enough trouble enforcing the 55 mph speed limit without this song, sales went up to 2,500." When WPLO radio in Atlanta staged their own convoy with Fries at the front door, the FCC sent a small army of investigators down to capture unlicensed CBers. "There were more than 1,200 vehicles on the interstate and bears in the air and Smokies everywhere, just like on the song," said Fries, "but the thing turned out pretty good. WPLO set up booths and a lot of CBers actually filled out license applications."

A talented scenic designer, adman, art director, and writer, Fries was never a truck driver, but he did grow up around trucks and trucks stops, and has had a CB in his jeep for four years.

"Mom and Dad had a business shipping prefabricated farm buildings around the country on flat-bed semis and, as a kid, I used to ride along, he recalls. "For awhile I had a job painting trucks, but what I really am is a writer, and you know any writer worth his salt does a lot of research. That's what I did and it sure turned out beautiful. "Convoy" has literally picked up my life and turned it around."

Jay Huguely (Cledus Maggard) didn't do too much research before he wrote his CB hit single, "The White Knight," the story of a devious Smokey who entraps speeding truckers by pretending to be offering his services as the "front door" in a convoy. A serious actor and director who studied at the Royal Academy in London and has taught Shakespeare, Huguely had been working in advertising for about two years when his boss, Bill Leslie, suggested he go out and listen to some CB transmissions and see if there was any material there for a TV commerical. After listening for no more than an hour, Huguely claims he rushed back to the studio and laid down the first version of "The White Knight." "I then got a couple of stock background music pieces out from our media file, pattered to it, and ad libbed the song"

Jay Huguely invented the Cledus Maggard character twenty years ago when he wrote a radio program in high school. He didn't decide to use it on the record until after the master tape had been cut and Mercury asked Huguely what name he wanted to use. "Cledus Maggard came right to mind; he'd always been a comical character and I thought it was perfect for the song. I come from eastern Kentucky where the name Cledus and the name Maggard are very common and I just put them together."

"The White Knight" was followed by Huguely's first album, *Cledus Maggard & The Citizen's Band,* a clever collection of comedy songs written by Huguely and Jerry Kennedy. An actor, singer, director, and writer (he wrote for the satirical TV show, "That Was The Week That Was") Huguely's 14-year career in show business is evident on the album. He does all the voices, more than a dozen of them, and his lyrics reflect his having previously written eight complete musical comedies.

There's a complete collection of CB themes on this album: a tragic-comic love song, "Sweet Mercy Day," where an abandoned husband gets on the CB and begs the wife he apparently cheated on to come on back home:

Come on, give me a break honey
Me and them beavers ain't been datin'
We just been doing a little modulating.

Then there's a song sung by a bedraggled father who loses his rocking chair spot in a convoy because his son keeps screaming, "Dad, I gotta go!" and the father has to take him to a Texaco station. There's a CB 1950s nostalgia song about a dream Cledus Maggard had 'bout having a CB back in the Wildroot, Annette Funicello days. "Jaw Jacking," is about all the nonsensical chatter that often fills the airwaves, and, of course, there's a bicentennial fantasy, "CB '76," where Huguely lets his imagination run wild and gives our country's heros CB sets. "Channel 1, if by land. Channel 2, if by sea," says Paul Revere.

There's no doubt about it. The hits of Fries and Huguely, the power of the Big Juicers, and the recent mythologizing of the trucker have overwhelmed the public with a desire to get their ears and start modulating. And that's what's been happening. From hookers to housewives, from hitchhikers to businessmen, hundreds of thousands, indeed, millions of new CBers are discovering the thrill of "going on the air" and exploring CB land.

Let's Truck Together

"Let's Truck Together," one of the first CB songs, was written by Kenny Price and his wife Donna. The tune was about two years ahead of its time when it first appeared in 1974, but nevertheless became a truckers' favorite.

I gotta rig and she's plenty big
And I ride the ribbons of gray.
But at fifty-five miles an hour
It's hard to make a livin' today.
So I took a little cash and mount it to the dash
With a CB radio.
I gotta the rabbit ears flappin' in the wind tonight
And talk to everybody I know.

I'll shake the trees
And you'll rake the leaves
And we'll truck 'em down the highway.
Not a Smokey bear in sight, put the hammer down tight;
Let's truck together.

All my buddies got a handle.
So we don't have to give 'em our names.
There's Tennessee Streaker and the Alabama Wildman,
Big Wheels and Jessie James.
What we're saying might sound like double talk,
But if you're rolling along, my friends,
Get yourself a CB if you want to truck with me
And tune in to channel ten.

1974 BMI, by Donna and Kenny Price

Me and Ole CB

All-time trucker favorite, Dave Dudley, wrote this with Ronnie Rogers in 1975. Although prophesying the imminent popularity of citizens band radio, "Me And Ole CB" became a favorite with the truckers' and country music fans, but never really "crossed over" to the pop charts.

I was ridin' this eighteen wheeler before they painted lines
Never made much money 'cause I couldn't make no time
Use to be a runnin' battle between ole Smokey and me

But now I got a six foot Shakespeare me and ole CB

Tonight I got the front door, can't always have that rockin' chair
Pig Pen's got the back door closed, just told me on the air
I just checked out the west bound, Road Hog says she's clean
Looks like a money-makin' run for me and ole CB

Ace, the blue eyed Indian, still runs with single drive
Ole Single drives a rachet jaw, but he keeps the night alive
Jolly Jack, the Beaver Nut, checking covers on the seat
We're looking for a red Pinto, me and ole CB

There goes Circle City Peddler, boy, he's getting fat
Makes his turn at Derby Town, I couldn't live like that
The chicken coop is open and so is DOT
'Preciate that info, Cisco, me and ole CB

California Devil down at marker twenty-five
He just hollered 'Breaker, the bushes are alive
Comb your hair, they're takin' pictures down at exit twenty-three,'
We'll give 'em a big 516 smile, me and ole CB

Look at that four wheel bear bait, just blew off my back door
Ole Smokey's gonna get him, that's a big ten four
Hawk Eye's had the hammer down since Nashville, Tennessee
He just saved some green stamps, thanks to me and ole CB

Eight Ball from Shakey City, he's talkin' to a base
Comin' on to Kewpie Doll, you sound like a pretty face
Wish he'd give me a breaker. I'd tell her to go to three
She might be runnin' Single, like me and ole CB

Windy City Charlie, he's down at marker nine
I just checked my twenty and I ain't runnin' far behind
Me and the eighteen wheeler were just about home free
Made damn good time, didn't pay one fine, me and ole CB

Convoy

The first true CB hit, "Convoy," "crossed-over" and took the number one spot on both the country and pop charts early in 1976. A platinum record in the US, it was number one in Australia, South Africa and Sweden and climbed near to the top of the charts in England, West Germany, and Japan.

Uh, breaker one-nine, this here's the Rubber Duck
You got a copy on me Pig-Pen, c'mon?
Uh, yeah 10-4, Pig-Pen, fer sure, fer sure
By golly, it's clean clear to Flag Town, c'mon?
Yeah, that's a big 10-4 there Pig-Pen, yeah
We definitely got the front door good buddy
Mercy sakes alive, looks like we got us a convoy

Was the dark of the moon on the sixth of June
In a Kenworth pulling logs
Cabover Pete with a reefer on
An a Jimmy haulin' hogs

We's heading fer bear on I-one-oh
'Bout a mile out a' Shakey Town

I sez Pig-Pen, this here's th' Rubber Duck
An' I'm about to put the hammer down

Uh, breaker Pig-Pen, this here's th' Duck
An' a-you wanna back off with them hogs

10-4, 'bout five miles or so, 10-roger
Them hogs is gettin' in-tense up here

By the time we got into Tulsa Town
We had eighty-five trucks in all
But they's a road block up on the clover leaf
An' them bears was wall to wall

Yeah, them Smokies as thick as bugs on a bumper
They even had a bear in the air
I sez callin' all trucks,
We about to go a-hunting bear

Uh, you wanna gimme a 10-9 on that Pig-Pen?
Uh, negatory Pig-Pen yer still too close
Yeah, them hogs is startin' to close up my sinuses
Mercy sakes you better back off another ten

Well, we rolled up interstate fourty-four
Like a rocket sled on rails

We tore up all our swindle sheets
And left 'em settin' on the scales

By the time we hit that Chi-town
Them bears was-a-gettin' smart
They'd brought up some reinforcements
From the Illinois National Guard

There was armored cars and tanks and jeeps
An' rigs of every size
Yeah, them chicken coops was full o' bears
And choppers filled the skies

Well, we laid a strip for the Jersey shore
And prepared to cross the line
I could see the bridge was lined with bears
But I didn't have a doggone dime

I sez Pig-Pen this here's the Rubber Duck
We just ain't goin' pay no toll
So we crashed the gate doin' ninety-eight
I sez let them truckers roll, 10-4

The White Knight

Released shortly after "Convoy," Cledus Maggard's "The White Knight" shows what happens when you have the misfortune of finding yourself with a Smokey for your front door. The tune ran strong on both the country and pop charts throughout 1976.

Now at seventy-five, or eighty-five, or I-twenty thuther way
Turn your squelch to the right and in the night

You'll hear some good buddy say
Breaker, Breaker, we've got a picture taker
Ole Smokey's at forty-three
It's that Japanese toy, that trucker's joy
That everybody calls CB, yeah, Citizen's Band
Keeps you up to date with the fender benders
And the Tijuana taxis, and them bears out there
A' flip floppin'

Now ahead of your children, and ahead of your wife
On the list of the ten best things in life
Your CB's gotta rate right around number four
'Course, Beavers, Hot Biscuits, and Merle Haggard
Come 1, 2, 3, you know

Well, I's loaded down, comin' out of Lake City
I's checking out seat covers, all young and pretty
When all of a sudden, there come a call over my CB
Ringing wall to wall
Said go to Double Nickels, as you hit the ridge
'Cause there's a smokey picture taker, t'other side of the bridge

Oh mercy, 'preciate that, good buddy
Ah, what's your handle there, come on
Got any county mounties out there a prowlin', come on
And he said: Ten four, Back Door
Put the pedal to the metal and let it roar
Hammer down to Macon Town, gonna see my momma shore
Well, the bears are gone, let's bring it on
The Georgia line's out of sight
Pulled out of Richmond Town last Saturday night
And my handle is the White Knight, how 'bout it

All right, White Knight, hammer down, you got the Mean Machine here
Well there I was a streakin', my needle was a peakin'
Right around seventy-nine
That ole diesel juice was a gettin' loose and everything was fine
When wall to wall, I got a call from a front door big bear trapper
Said break one nine, good buddy of mine
You got a Smokey in a plain white wrapper
Well, I jammed my stick, I lost twenty-eight quick
You could hear them gears a tearin'
Little gravels in my wheels going ping, ping, ping

Hey there Smokey ole buddy, tell me if I'm right
Are you my front door, are you the White Knight, come on
And he said: Ten four, Back Door, you in a heap of trouble, boy,
For sure, gonna read you your rights, and treat you fair
Just pull over there with your rocking chair
Want you boys to know each other real well
'Cause you gonna be sharing the same jail cell
You make twelve cotton pickers I've caught tonight
Runnin' front door as that White Knight, (laughter)

Listening In On the Party Line: Some Diverse Users of Citizens Band Radio

On truck mirrors and car trunks, on highways across the country, on apartment balconies, and farm house rooftops they can be seen—thin, metal stalks whipping in the wind, signposts of a new age in two-way communications that is as good as its name—citizen's band.

Motorists are the most obvious users—CB's short-range service is ideal for the exchange of road and traffic information. And even though some states contend it's illegal, ("obstructing justice"), many drivers use CB to alert others to the presence of the highway patrol. Farmers use it to stay in touch while they're in the fields. Commercial services use it to dispatch their cars and trucks. The possibilities for using CB, in fact, are limitless and will depend upon your occupation (salesperson, trucker, service person, doctor, lawyer, private detective, executive, plumber, rancher), leisure-time activities (traveling, hunting, fishing, boating, camping), or location (city, country, small town, desert, ocean). You can use CB to provide communication between your car and your home or office; or between your home and office. Use CB to call home when you're stuck in traffic and late for dinner; call ahead on your way home from work to keep in touch with the baby sitter or with a neighbor while you're out shopping—use CB to remind your husband or wife you need extra groceries or to pick up the laundry.

Two-way CB communications are used by forestry services, ski patrols, TV technicians, surveyors, electricians, construction crews, security patrols, highway maintenance crews, business executives, truckers, farmers, and salespeople. Boaters use CB for low-cost ship to shore communications. Auto racing teams use it to communicate between driver and pit crew.

In many areas, you can use your CB radio to call ahead and make reservations at hotels, motels and restaurants when you're out in the highway in your car. You can even call CB-equipped garages or service stations to obtain assistance if you're having car trouble. Of course, you can also utilize CB's Channel 9 national emergency facility — literally tens of thousands of highways are monitored on Channel 9 by public assistance organizations, individuals, police departments, rescue units, hospitals, garages and others who can provide aid or simply street directions, in the event of an emergency. As you begin listening in on the CB party line, though, you'll begin to see for

yourself some of the diverse uses to which CB can be put.

Once you're on the party line, you'll be able to modulate with so many different people. One of the more enterprising CBers you're likely to meet is Lulubell, the pleasant, matronly madam of a flourishing whorehouse in Wheeling, West Virginia. What makes Lulubell unique, however, is the fact that she was the first of her profession to go entirely CB.

Located just a few miles down I-80 from the famous Windmill Truck Stop (itself no mean purveyor of pleasures), Lulubell's is known throughout the Mid-West. In fact, it's not easy to pass by Wheeling without catching Lulubell herself on Channel 9. Her prim, high-pitched voice usually answers if you break with "a collect call for Lulubell" (Don't try to raise her much before three in the afternoon). Once you've established contact, that motherly, Minnie Pearl voice will lead you off exit 59 and through a labyrinth of local streets until you're parked in front of an innocent, white-framed Victorian triple decker, wrap-around front porch and all. The street is so quiet you may wonder whether you're in the right place. Lulubell's, by the way, is not one of your "Professional Truckers Only" establishments, like those notorious joints along South Carolina's Alternate Route 17. With a clientele composed of over-the-road haulers, traveling salesmen, college men, and local businessmen and politicians, Lulubell decided long ago to travel first class. She checks out all of her clients by CB and swears that it's possible to weed out the rednecks and roughhousers with just a few minutes' modulation. Most of the local CBers have known about Lulubell for years and they help screen the clientele—if you sound like an undesirable, you simply will never get a response to your break. But back at the house, Lulubell's is all business—CB business! You can tell that by the 60-ft maximum-height Moonraker antenna that sprouts from the rear second-floor sleeping porch. Just inside the front door is Lulubell's radio room, fitted out with a Browning Golden Eagle base station and a bank of four telephones (just in case business gets heavy). The Highway Patrol in this part of West Virginia generally have CB units in their cars, as do most of the local Wheeling Police, so Lulubell's existence is no secret to her neighbors. Yet, no one really seems to mind, so orderly a house does she run and so efficient a system of communication does she maintain. About the only real complaints anyone ever hears about Lulubell's crop up each September during WWVA's annual Trucker's Jamboree and Convention, when disgruntled gear jammers, who've traveled hundreds of miles to see the new Petes and K-Whoppers, learn that seeing is all they can do—test driving is out. These fellows invariably wind their way to Lulubell's and not even this little old lady of the airwaves can keep them quiet. With a half dozen or more 18-wheelers double-parked on the street outside, and a good deal of romping and stomping going on inside, Lulubell's usually finds it politic to close down for the next few weeks. "Besides," says the Madam herself, "it wouldn't do to stay open during the elections."

If Lulubell was a first, she was soon overtaken in the field of CB assignation. Today, motorist's along virtually any interstate can pry loose from local ratchet jaws the handle of the local madam ("Jolie, the Cajun Queen" reigns supreme in New Orleans; the "Black Widow" purveys pleasure to the Miami area; and, just across the river from our nation's Capital, "Rebel Gal" hovers around Channel 15, ready to guide you to her place of business in Arlington, Virginia.)

Not to be outdone by these "base-station operators," a new generation of mobile CB hookers has taken to the highways. Preying primarily on truckers, they'll usually take on anybody they find on the air. Some, like "Golden Girl" in Bakersfield, will ride right up your rear bumper before you realize what you've got yourself into. Others prefer to stay out of sight (or at least unidentified) until all the arrangements are made ("Meet you in 15 minutes at the Day's Inn, Exit 5-9, room 1-0, 25 green stamps").

These highway hookers have certainly livened up more than a few monotonous journeys. But beware the come-hither voice that floats out over the CB from a roadside base station. The chances are that sexy-sounding beaver telling you how great you sound, cracking risque jokes as fast as the mile-markers flip by, is just a frustrated and probably frumpy farmer's daughter (or sister, or wife) getting off on this new little toy she's found called CB. One young lady in Gary, Indiana, who went too far was shocked late one night to see an 18-wheeler pull right up in her driveway, even though she lived some 5 miles off the Interstate. What she said to her father, or the trucker, has never been recorded.

A teenager in Salt Lake City, who had watched CB work effectively out on the Bonneville Salt Flats (he had worked the catering service during a successful world record speed attempt by the Harry Miller and Larry Lange-Kenworth tractor), introduced another innovative application for CB. A regular hitchhiker between Salt Lake and his college on the Coast, Larry Tompkins was seized with inspiration on his way home last Christmas. Instead of hiding out to the entrance ramp of I-95 and hoping his well-scrubbed appearance and neatly worded sign would get him a quick ride, Larry lay back at the Holiday Inn coffee shop and pulled out a 4-channel CB walkie talkie. Within minutes, he had his ride (after he had rejected another that would have taken him only half way home) and knew quite a bit about his driver as well. A few weeks later, Larry was back in Southern California organizing a massive dispatch service for hitch-hikers all over the Far West. For a nominal fee, he would pair a hiker with a ride East, with a minimum of waiting and very nearly door to door transportation.Tompkins' monopoly on CB hitch-hiking ended, however, when walkie talkies started sprouting from the back packs of hundreds of highway hustlers. Today, the old-fashioned hitch-hiker without ears is a rare sight indeed, and those who aren't on CB themselves have enough good sense to hitch rides with motorists who are, so that their next ride can be negotiated before they leave the first one.

Obviously many Holiday Inns and Howard Johnson's boast maximum strength 60-ft CB antennas these days. They were among the first to realize the business boom inherent in a short-notice "CB reservations" system. Many lesser-known establishments, once written off because they had been bypassed by the Interstate have also prospered, thanks to CB. Anyone looking for a choke and chew or a place to go bowling just has to get on 11 (the national contact channel) or 19 (the trucker's channel) to find out directions to the best spots in town.

When you're traveling through parts unknown, it's a good bet the voice answering your call will come from a local housewife monitoring on her new base station. Literally thousands of homebound housewives have livened their days by listening for hours on end to the incessant squawking of their treasured CBs. It's a fact that most of the serious CB biscuit burners are women —and they are usually the first to suggest organizing a local CB club or participating in a CB coffee break. And what woman, left alone or with only small children, wouldn't feel more secure knowing that help in any emergency was just minutes away on Channel 9.

Not everybody on the CB airwaves, however, is welcomed with open arms. Down Alligator Alley (Florida) along I-75, several truckers who should know say, "The highway patrol are notorious for using their two-ways in illegal entrapment of speeders. Just like Cledus Maggard's 'White Knight,' the Floriday troopers would give you the all-clear signal over Channel 19, then sit there waiting for you to roll right into their arms as you come over the hill. We finally quit talking to anyone around there unless we recognized his voice." What about Smokey's side of the story? Surprisingly, there is a growing number of law enforcement officials who condone CB. The superintendent of the Missouri Highway Patrol, Samuel S. Smith, for example, says, "CB in trucks and cars is the greatest thing to come down the pike since the invention of the fifth wheel. When the motorists began to notice we had ears, they began reporting intoxicated drivers, wrong-way drivers, stranded motorists, accidents, and other matters requiring law enforcement action." A Wyoming highway patrolman puts it more bluntly: "Sure, there's a lot of chatter on CB, but I'll listen to that for a week just to get a report of one accident or one drunk driver. As for speed . . . if CB helps slow people down even part of the time, it helps us do our job and cuts down the number of tickets we have to write."

One of the unexpected benefits of CB is the tremendous diversity of the individuals you'll meet out there on the airwaves. Unlike ham radio, where every operator is equipment-proud and more than a little obsessed with DXing, CB is truly the citizen's band, and that invisible nation of the airwaves is made of people from all walks of life. Muhammed Ali, for example, has been a CB buff for several years and whiles away the many hours he must spend in his bus (or his limosines) by carrying on a nonstop dialogue with motorists who have recognized the "Big Bopper." Ali, in fact, is a favorite of truckers on the Pennsylvania and Jersey turnpikes and will often give blow by blow accounts of his latest fight to gear jamming CBers who weren't able to catch the films on television.

Senator Barry Goldwater, long an amateur radio ticket holder, has also held a CB license for several years. The Senator uses his CB rig primarily on the boat which he keeps in California, but he's keenly aware of both the problems and potential of citizens band broadcasting and hopes legitimate CBers will help police their service and cut down the problem of unlicensed and illegal broadcasters before the FCC has to step in and impose stricter regulations. Most rock and country musicians who tour have become familiar with the virtues of citizens band, but Johnny Cash is one CBer who never has to use a handle. His unmistakable, booming voice is instantly recognizable and he professes enjoying "nattering with the fans" out there in CB land. Even comedienne/actress Lily Tomlin has allowed as how CB has become an integral part of her life on the road. Without having to resort to her "Ernestine, The Operator" character, Lily says that CB has saved the day more than once "when traveling difficulties made it appear possible that the show simply wouldn't go on."

Perhaps the most unusual adherent of CB so far, however, is the anonymous sixth place finisher in last year's Fourth Annual Cannonball Baker Sea To Shining Sea Memorial Trophy Dash. Driving a factory-standard Dodge Van equipped with a Johnson Messenger III transceiver, he crossed the country (with two co-drivers) in 35 hours and 12 minutes for an almost unbelievable average of 90.3 mph. "We made incredible time and only got three tickets, two of them in Kansas," he reported, "and we never could have done anything like this without our CB."

3.

CB Talk: A Wall-to-Wall Guide to the Language of Citizens Band Radio

"It's a whole new world. . . . I was on my new CB for five full hours before anyone even would give me the time of day."

A RUBBERBANDER

You've Got to Learn the Language

A CB radio and an FCC license will get you over the border, but if you want to be a listened-to and talked-to citizen of CB land, you've got to learn the language. With over 500,000 new CBers immigrating into the airwaves each month, the channels around the big cities and along the oil-stained super slabs have become jammed with rubberbanders and ratchet jaws. And although most veteran CBers are generally friendly and helpful, they haven't got the time or inclination to talk with neophyte 4-wheelers who want to try out their new rigs.

Fortunately with a little common sense and practice anyone can learn to sound like the driver of an 18-wheel K-Whopper, and get talked to on the air waves. You can be commuting in your pregnant roller skate or camping in your living room and still sound like the real McCoy. That's one of the great attractions of CB land. You can be anyone you can *sound* like. You'll find yourself acting out all kinds of fantasies as you modulate behind a mask of anonymity. And before you know it, your true CB self will emerge.

The language of CB land is an eclectic collection of truckers' lingo, ham-radio slang, military, police, and aviation jargon, hippie talk and just plain (and often poetic) bad English. In radio's earliest days, there was no voice transmission — Morse code was laboriously pounded out a dot and a dash at a time. To shorten messages, numbered codes were employed instead of words: a "73" meant "best regards" and an "88" was the code female ham operators used for "love and kisses." CBers adopted these number codes, adding them to their own colorful colloquialisms. Today you'll hear: "3s and 8s" going out to good buddy cottonpickers all over the country.

The traditional Ten Code has almost one hundred entries, many of which are inappropriate and unnecessary for today's CBers (eg, 10-99, "mission completed, all units secure.") A more up to date Ten Signals List with only 34 entries was introduced by APCO (Associated Public Communications Officers) and revised in 1974. This new Ten Code has caused confusion among older CBers, since the code for a word like "Mayday," for example, has been changed to mean a simple request for a time check (10-34). However, the APCO code is fast becoming the preference among CBers.

Of course, the joy of talking CB has little to do with efficiency of expression. CBers "get their signals off" from the play of language, alliteration, rhythm and rhyme. It's fair to say that they're among the most creative broadcasters on the air. Just put your ears on and listen: "Shake the trees and rake the leaves," means keep an eye out for "bears" (police) who may be hiding "in the bushes" (off the road). When a CBer spots the law, he'll tell other mobiles to "brush their teeth and comb their hair" (be on their best behavior). A patrolman with a radar trap is a "Kojak with a Kodak." A small-town patrolman is a "local yokel," while a sheriff's deputy becomes a "county mounty." A police car with flashing lights and siren on top is a "Tijuana taxi." When it's safe to accelerate, you'll hear, "put the pedal to the metal," or "put the hammer down; it's clean and green ahead." "Keep the shiny side up and the greasy side down," means accelerate, and when you're driving along the West Coast you're on the "Shakey Side." When you're in the "Bikini State," "eyeballing" some "seat covers," you're actually in Florida, looking at the girls.

Communicating, though is more than knowing the language. CB talk is half words and half pronunciation. If you've got some country in your soul or twang in your talk, you shouldn't have trouble finding other cottonpickers to talk to. But if you're a city slicker, prepare for a real personality change if you want to get anywhere in CB land. City slang needs to be changed into a countrified, cowboyesque, truck drivin' drawl; a "10-fer" as often as a "10-4" and a hearty "come-awn, come-awn ald buddy, anyone out there?" Though jokes are made about new CBers spending all their time trying to get a time check, the truth is that once you've got your first 10-34, you're off to a good start and probably ready to move on to more meaningful conversation.

Once regionalized and parochial, CB talk has now become a national language. No matter where you are in the United States, the CB vocabulary is basically the same, even though certain words may be more popular in their area of origin. "Cowboy," for example, was once a somewhat derogatory term used along the West Coast to denote stud truckers who spent more time polishing their chrome than driving their rig. Now a dude anywhere is likely to be called a cowboy. Though the language has become universal, speaking patterns still vary. Many truckers report that as you roll into the West, the talk gets more businesslike and dignified, though by the time you hit the Shakey Side, it gets crazy. In the South there's a lot of "nonsensical, slow drawlin with macho ratchet jawing jes' 'bout nothing," while back in New England, conversation is minimal. There's lots of chatter and rubberbanding on the Dirty Side (the Boston-New York Washington run), although the truckers' are surprisingly quiet. Truckers seem to absorb the mean, distrustful, alienated East Coast feeling and aren't willing to extend themselves, even on CB. As one roadmaster puts it: "The pollution is everywhere, even on the airwaves, and nobody wants to open his mouth." Around West Virginia, the talk gets pretty raunchy. Pavement princesses hover around the larger truck stops (hooker havens), filling the airwaves with seductive solicitations. Some of the best states for freewheeling chatter are Pennsylvania, Ohio, Indiana, and Missouri. And, of course, the best time to tune in is night time, when truckers these days tell stories instead of popping pills to keep alert and awake.

If you're ready to enter CB land but suffer from "mike fright," relax. Start by monitoring a few minutes. Let some of that good

CB feeling run through your brain; have a cup of 30-weight (coffee) and spend time learning the lingo. It's your passport to an invisible nation with adventure just around the channel.

Note: Electronic and other technical terms are not included in this CB Dictionary, but are included in the Technical Glossary in the Appendix.

CB Dictionary

Alligator

ACE a CBer with a powerful transceiver, big antenna, and bigger ego, who thinks of himself as the number one channel master.

ADIOS goodbye, fairwell, finished talking.

ADVERTISING flashing lights, antennas, numbers, etc., on a marked police car.

AFFIRMATIVE yes.

ALLIGATOR 1. a CBer who makes a transmission, but doesn't reply, "That cottonpickin' alligator is all mouth and no ears." 2. a linear amplifier.

ALLIGATOR ALLEY Florida.

ANCHOR MAN a base-station operator.

ANCHORED MODULATOR same as ANCHOR MAN.

APPLE a CB addict, "He's an apple to the core."

ASTRODOME CITY Houston, Texas.

BABY BEAR rookie policeman.

BABY BUGGY school bus.

BACK replying back.

BACKDOOR 1. last CB vehicle in a line of two or more, see CONVOY. 2. the road behind.

BACK DOOR CLOSED rear of convoy covered for police.

BACK DOOR SEALED UP same as above.

BACKGROUND interference noise heard on CB radio.

BACK OFF 1. stop transmitting. 2. slow down.

BACK OFF ON IT slow down.

BACK OFF THE HAMMER slow down.

BACK ON DOWN 1. slow down. 2. stop transmitting.

BACK 'EM UP same as above.

BACK 'EM ON OUT finish talking.

BACKSIDE return trip. "We's got the hammer down on the backside and we're heading for home-20."

BACK TO YOU used to let whomever you're talking to know it's his turn to talk, also COME BACK, COME ON, TAKE IT BACK, BRING IT BACK.

BACKSTROKE return trip.

BAD SCENE cluttered CB channel.

BAGGING THEM police arresting speeders.

BALLET DANCER CB antenna blowing in the wind, especially a center or top-loaded antenna.

BALLOON FREIGHT 1. light load. 2. load of household furniture.

BALLOON TIRES radial tires.

BALONEYS tires.

BAND BENDER single sideband user.

BANDMASTER 1. person acting as a "dispatcher" who gives others permission to talk on a channel. 2. person with a very powerful signal.

BANGING 1. hunting. 2. looking for.

BAO BAB extra-wide load, (from Africa's Bao bab tree).

BAREFOOT operating a CB within legal power limit. "I'm running barefoot."

BAREFOOT MOBILE mobile CB rig with no extra power.
BARLEY POP beer.
BARN truck garage.
BAR CITY Forrest City, Arkansas.
BARNIES police.
BASEMENT Channel 1.
BASE STATION a transceiver maintained at a fixed location.
BAY CITY San Francisco, California.
BEAM a highly directional antenna.
BEAN STORE a restaurant.
BEAN WAGON coffee wagon.
BEAR police of any kind.
BEAR BAIT a speeding vehicle without CB.
BEAR BITE speeding ticket.
BEAR CAGE police station.
BEAR CAVE police or highway patrol station located on a highway.
BEAR DEN police station located off the highway.
BEAR FOOD speeding vehicle without CB.
BEAR IN THE AIR police helicopter or airplane.
BEAR IN THE SKY same as BEAR IN THE AIR.
BEAR MEAT same as BEAR FOOD.
BEARDED BUDDY police of any kind.
BEARS ARE CRAWLING police are switching from one side of the road to the other.
BEAR REPORT request for information on police location.
BEAR SITUATION same as BEAR REPORT.
BEAR STORY police location report.
BEAT THE BUSHES first truck in a line of two or more is said to BEAT THE BUSHES as he checks out road conditions and location of police ahead, see CONVOY.
BEAVER any female.
BEAVER HUNT looking for girls.
BEAVER PATROL same as BEAVER HUNT.
BEAST a top quality CB rig.
BED BUG HAULER furniture mover.
BEER CITY Milwaukee, Wisconsin
BEING SOUTH same as heading south.
BELLY UP truck that's flipped over.
BENDING THE WINDOWS transmitting powerfully; "You're bending my windows."
BENNY CHASER coffee.
BETTER COOL IT slow down.
BETTER HALF wife.
BETWEEN THE SHEETS sleeping.
BIG A 1. Amarillo, Texas. 2. Atlanta, Georgia.
BIG BROTHER police of any kind.
BIG EARS clear reception.
BIG APPLE New York City, (also DEFAULT CITY).
BIG D. Dallas, Texas.
BIG DADDY the FCC.
BIG JUICER 1. all-night country music disc jockey. 2. AM radio station.
BIG M. Memphis, Tennessee.
BIG ORANGE a Snyder truck.
BIG R. a Roadway Freight System truck.
BIG SLAB expressway, (also SUPER SLAB).

Big Apple

BIG STRAPPER extremely powerful CB transmitter.

BIG SWITCH the on/off switch on a CB.

BIG T. Tucson, Arizona.

BIG 10-4 absolutely, "I'm with you 100%."

BIKINI STATE Florida, (also ALLIGATOR ALLEY).

BILLY GOAT an old timer, a CBer who ratchet jaws.

BIRD 1. Ford Thunderbird. 2. Pontiac Firebird.

BIRD IN THE AIR police helicopter.

BISCUIT BURNER 1. old timer CBer. 2. CB club member.

BIT ON THE SEAT OF THE BRITCHES got a speeding ticket.

BLACK AND WHITE CBer police car with CB.

BLACKLIST violations recorded on driver's license.

BLACK WATER coffee, (also THIRTY-WEIGHT, MUD, et al).

BLEEDING signal from adjacent channel interrupting transmission.

BLEED OVER same as BLEEDING.

BLEED WORTHY likely to bleed over, "Using a linear will make you bleed worthy."

BLESSED EVENT a new CB rig.

BLEW MY DOORS OFF passed me with great speed.

BLINKIN' WINKIN' school bus.

BLOCKING THE CHANNEL pressing the microphone switch without talking thereby preventing others from communicating.

BLOWING SMOKE a loud, strong signal. "You're blowing smoke tonight, good buddy."

Blowing Smoke

BLOOD BANK ambulance.

BLOOD BOX same as above.

BLEEDING BLOOD BANK ambulance carrying injured passenger.

BLUE AND WHITE police.

BLUE BOOK CB dictionary.

BLUE BOYS police.

BLUE JEANS state troopers.

BLUE LIGHT MACHINE police car with flashing light.

BLUE SKIES AND GREEN LIGHTS TO YOU best wishes.

BLUE SLIP speeding ticket.

BOASTIE TOASTIE a CB expert.

BOB TAILING a truck tractor traveling without a trailer.

BODACIOUS powerful, super strong.

BOOGIE MAN state trooper.

BOONDOCK taking back roads to avoid a weigh station.

BOOTS linear amplifier or other illegal signal booster.

BOULEVARD interstate highway or road with limited access.

BOUNCE AROUND 1. return trip. 2. next trip through.

BOUNCING CARDBOARD driver's license.

BOUND AROUND same as BOUNCE AROUND.

BOZOS nighttime CBers especially in Nashville, Tennessee.

BOX 1. tractor trailer. 2. CB set.

BOX ON WHEELS hearse.

BOY SCOUTS state police.

BRA BUSTER bosomy woman.

BRASS POUNDER amateur or ham operator.

BREAK or BREAK BREAK used to ask permission to enter a channel on which there is conversation. (Usually used with the channel number on which sender wishes to speak.)

BREAK a meeting of local CBers.(Also, COFFEE BREAK.)

BREAKER a CBer who wants to come in on a channel, "Go ahead, BREAKER."

BREAKER BROKE variation of BREAKER.

BREAK FOR (SPECIFIC PERSON) a call for a specific CBer on the channel.

BREAKING THE OLD NEEDLE transmitting powerfully.

BREAKING UP signal is erratic and difficult to comprehend.

BREAKING WIND first vehicle in a line of two or more.

BREEZE IT forget it.

BRING IT BACK answer back.

BRING YOURSELF ON IT 1. answer back. 2. move over to the right lane.

BROWN BOTTLES beer.

BROWN BOTTLE CITY Milwaukee, Wisconsin.

BROWN PAPER BAG unmarked police car.

BRUSH YOUR TEETH AND COMB YOUR HAIR be on your best behavior, police radar ahead. "Brush your teeth and comb your hair, there's a Smokey bear in the bushes."

B. TOWN Birmingham, Alabama.

BTO big-time CB operator.

BUBBLE GUM MACHINE any vehicle with flashing lights on top, usually a police car.

BUBBLE GUMMER teenage CBer.

BUBBLE TROUBLE tire problem.

BUCK ROGERS MACHINE a hand-held radar unit used by certain highway patrols.

BUCKET MOUTH a dirty talker who pollutes the airwaves.

BUCKET OF BOLTS tractor-trailer rig, (also any truck).

BUCKEYE STATE Ohio.

BUDDY fellow trucker or CBer.

BUGGER HOLE BUNCH Irvine, California.

BUG OUT to leave a channel.

BUGS ON THE GLASS insects on the windshield.

BULL RACK a truck hauling meat.

BULL WAGON a truck hauling live cattle.

BULLET LANE passing lane.

BULL JOCKEY an idle talker.

BULLDOG Mack truck.

BULLS police.

BUMPER LANE the passing lane.

BUTCHERBOY driver of meat truck.

BUTTON PUSHER a person who pushes his microphone button without talking, thereby causing interference and/or preventing others from using the channel.

BURYING YOUR HEAD IN THE BOOK learning Part 95 of the FCC rules and regulations governing citizens band radio.

BUY AN ORCHARD a highway accident.

BUSHEL 1000 pounds.

BY listening on the channel, awaiting calls.

COE cab over engine, (cf. LONG NOSE).

CACTUS JUICE liquor.

CACTUS PATCH Phoenix, Arizona, or Roswell, New Mexico.

CALIFORNIA GRAPEVINE U.S. 1 between L.A. and San Francisco.

CALIFORNIA LIGHTS silver, bullet-shaped clearance lights.

CALL LETTERS official FCC letters and numbers assigned to a CB operator.

CAMERA police radar, (also Kodak). "There's a bear with a camera at Exit 20."

CANDY MAN FCC field enforcement officer.

Cactus Patch

Camera

Cattle Car.

CAPITAL J. Jackson, Mississippi.
CARRIER a signal transmitted without modulation, see KEYING THE MIKE.
CARTEL a group hogging a channel.
CASA home, (also home-20).
CATCH CAR police car used to make speeding arrests, while another policemen is manning radar.
CATTLE CAR bus. (Careless cattle car is a Greyhound bus.)
CB citizens band radio.
CB BIBLE the Final Word on CB.
CELL BLOCK location of a base station.
CB LAND network of CBers, the "overground underground of the airwaves."
CHAIN GANG members of a CB club.
CHANNEL MASTER person acting as a "dispatcher," who gives permission to talk on a channel.
CHARLIE, CHARLIE-CHARLIE, or CHARLIE BROWN yes.
CHASE CAR 1. police car with hidden radar. 2. a catch car, (see above).
CHECKING YOUR EYELIDS FOR PINHOLES a state of mind characterized by extreme fatigue.
CHECK THE SEAT COVERS look at the (female) occupants in a car.
CHICK woman, girl.
CHICKEN CHOKER poultry truck.
CHICKEN COOP weigh station.
CHICKEN COOP IS CLEAN weigh station is closed.
CHICKEN INSPECTOR weigh station inspector.
CHICKEN PLUCKER ICC weigh station officer.
CHOKE AND PUKE truck-stop restaurant, (also CHOKE and CHEW).
CHOKING THE CHICKEN avoiding a weigh station.
CHOO CHOO TOWN Chattanooga, Tennessee.
CHOPPED TOP 1. short antenna. 2. homosexual.
CHOPPER helicopter, usually with Smokies.
CHRISTMAS CARD speeding ticket.
CHRISTMAS TREE LIGHTS running lights around a truck cab.
CIGAR CITY Tampa, Florida.
CINDERELLA CITY Disneyland, California.
CINDERELLA WORLD Disneyworld, Florida.
CIRCLE CITY Indianapolis, Indiana.
CIRCUS WAGON Monfort truck.
CITY KITTY local police.
CLEAN AND GREEN road is clear of police and obstructions.
CLEAN AS A HOUND'S TOOTH same as above.
CLEAN CUT an unmodified truck, automobile, or CB rig.
CLEANER CHANNEL channel with less interference.
CLEAN SHOT no police or road construction visible.
CLEANING UP being sexually promiscuous.
CLEAR "I am through transmitting."
CLEAN AS A SPRING DAY road is clear of police.
CLEAR AFTER YOU WHEN YOU SIGN OFF the channel will be clear.
CLEAR AND ROLLING signing off and moving.
COAL BUCKET a coal-carrying truck.
COCKLE BURR pep pill.
COFFEE BREAK a local social gathering of CBers. (Also a CB club.)
COFFEE WESTERN STYLE strong coffee that's been on the range all day.
COLD COFFEE beer.
COLLECT CALL message for a specific CBer.

Colors Going Up

COLORADO KOOLAID Coors beer.
COLORS GOING UP policeman turning on lights atop patrol car.
COME BACK "answer me," also COME HERE, COME ON.
COMIC BOOKS truck driver's log book.
COMING IN LOUD AND PROUD strong, clear reception.
COMING OUT THE WINDOWS perfect reception.
CONCRETE JUNGLE expressway.
CONCRETE RIBBON same as above.
CONNIVIN' CANARY WOMAN female dispatcher.
CONVENTIONAL gas-powered engine.
CONVERSATE to talk.
CONVOY a line of trucks in CB contact. The first truck is the FRONT DOOR. He SHAKES THE TREES and lets the vehicles behind know if there are any police or radar traps. The last truck is the BACK DOOR. He RAKES THE LEAVES and reports any police coming up from behind. The vehicles in between the doors ride in relative security in the ROCKING-CHAIR position.
COOKIE BOX truck carrying bakery products.
COOKIES cigarettes, also WEEDS.
COOKING driving.
COOKING GOOD driving above the legal speed limit.
CONCENTRATOR driver.
COP FLOP police-car U turn.
COPY to understand or receive a transmission, "Do you copy?"
COPYING THE MAIL 1. listening in on a conversation. 2. receiving a clear signal.
CORN BINDER International truck.
CORN-FLAKE MACHINE Consolidated freight truck.
COTTON PICKER 1. a CBer you don't like. 2. an ironic term of endearment for a good fellow CBer. 3. catchall derogatory phrase.
COTTONTAIL female hitchhiker "with ears."
COUNTY MOUNTY local police or sheriff's deputy.
COUPON speeding ticket.
COUNTRY CADILLAC a pick-up truck.
COUNTRY JOE rural police.
COVER GIRL woman, as in SEAT COVER.
COVERED UP interfered with.
COW TOWN Fort Worth, Texas.
COWBOY CADILLAC 1. a Chevy El Camino or Ford Ranchero. 2. any pick-up truck.

Cradle Baby

CRADLE BABY a CBer who keeps trying to break, but lacks the guts to assert his right to enter the channel.
CRANK UP THE MIKE turn up the preamplifier.
CUB SCOUTS sheriff's deputies.
'CUDA Plymouth Barracuda.
CUP OF MUD coffee.
CUT LOOSE sign off, stop transmitting.
CUT THE COAX turn off the CB.
CUT SOME Z'S get some sleep.
DADDY-O the FCC.
DAGO San Diego, California.
DARK TIME night.
DEAD KEY 1. a person who presses a microphone switch without talking, causing interference. 2. a creep.
DEAD PEDAL 1. slow driver. 2. CB, RUBBERBANDER.
DECOY unmanned police car.

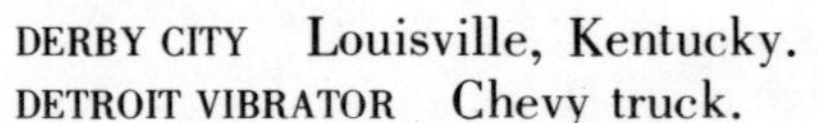

Dingy Weed

Dog House

DERBY CITY Louisville, Kentucky.

DETROIT VIBRATOR Chevy truck.

DIARRHEA OF THE MOUTH talking too much.

DICE CITY Las Vegas, Nevada.

DIESEL CAR a diesel-powered tractor.

DIESEL CITY Detroit, Michigan.

DIESEL DIGIT channel 19 truckers' channel.

DEFINITELY without a doubt (the most used adjective in CB lingo.)

DIG YOU OUT to understand, to pull in a signal through noise.

DINGY WEEDS the country.

DIVORCE CITY Las Vegas, Nevada.

DOG BOX gear box.

DOG HOUSE motor cover.

DOIN' IT THE OTHER WAY traveling in opposite direction.

DOIN' IT TO IT going full speed.

DO IT TO ME "answer me back," "come on."

DOING OUR THING IN THE LEFT-HAND LANE going full speed in the passing lane.

DOING THE 5-5 driving at 55 miles per hour.

DONKEY rear.

DON'T FEED THE BEARS don't get a speeding ticket.

DON'T LET YOUR TRICKING TRIP UP YOUR TRUCKING a sign off, "Drive safely."

DON'T LET YOUR TRUCKING TRIP UP YOUR TRICKING same as above, usually used together.

DOT Department of Transport, equivalent to FCC. Also Canadian Dep't. of Transport.

DOT MAN U.S. Dept. of Transport representive, who inspects trucks and gives tickets for violations.

DOUBLE E to see clearly with both eyes, to eyeball.

DOUBLE-BOTTOM RIGS trucks with twin trailers, articulated trailers.

DOUBLE-JOINTED CORN FLAKE BOX tractor with double (articulated) semi-trailers.

DOUBLE L telephone.

DOUBLE NICKEL the 55 mph speed limit.

DOUBLE SEVEN "no," "negative contact."

DOUBLE TRUCKERS twin, mirror-mounted CB antennas.

DOUGHNUT tire.

DOWN, I'M DOWN end of a transmission. "I'm signing off."

DOWN AND GONE a sign off, turning off CB.

DOWN AND ON THE SIDE through talking, but monitoring.

DOWNED stuck, mired in mud or snow.

DOWN ONE going to a lower channel.

DO YOU COPY? "Do you understand?"

DOZING stopped or parked.

DRAG OUT to pull in a weak signal.

DRAGGIN' WAGON a wrecker.

DRESS FOR SALE prostitute.

DROP THE HAMMER speed up, "Drop the hammer and eat up them white lines."

DROPPED IT OFF THE SHOULDER ran off the side of the highway.

DUDLEY DO-RIGHT a policeman.

DUMMY unmanned police car.

DUSTED MY BRITCHES 1. passed me. 2. transmitted at same time, blocking my signal.

DUSTED EARS transmission interrupted, "That cottonpicker dusted my

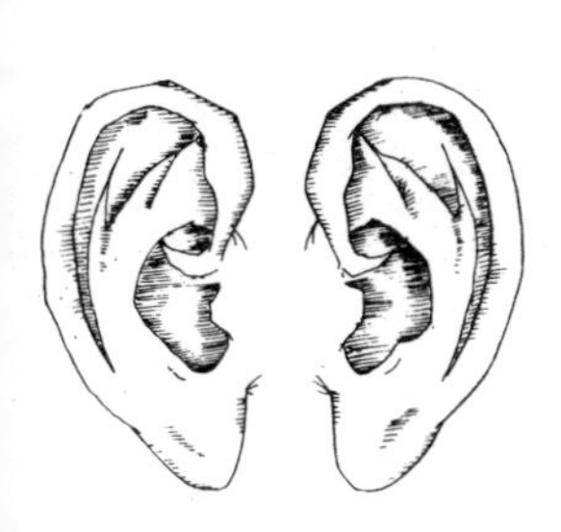

Ears

ears."

DX long distance. DXing means long-distance transmitting.

EARS 1. CB radio. 2. antennas.

EARS ON CB radio turned on.

EASY CHAIR CB vehicles in middle of a CB convoy, between the FRONT DOOR and the BACK DOOR, also ROCKING CHAIR.

EAT UP THEM WHITE LINES accelerate.

EATEM-UP roadside restaurant.

EIGHTEEN-LEGGED POGO STICK 18-wheel tractor-trailer truck.

EIGHTEEN WHEELER tractor-trailer truck.

EIGHTMILER four-wheeler who keeps you guessing for eight miles or more because he's left his turn signals on.

EIGHTS AND OTHER GOOD NUMBERS best wishes.

EIGHTY-EIGHTS love and kisses.

EIGHTY-EIGHTS AROUND THE HOUSE good luck and best to you and yours.

ELECTRIC TEETH police radar.

EVERYBODY MUST BE WALKING THE DOG all channels are busy.

Eyeball

EVEL KNEIVEL motorcycle rider.

EVEL KNEIVEL SMOKEY motorcycle police.

EXTRA MONEY TICKET speeding ticket.

EYE IN THE SKY police helicopter.

EYE BALL 1. meet a CBer face-to-face 2. to look at, "Put an eyeball on it."

EYE BALL TO EYE BALL two CBers together.

EYE BALL QUE-SO personal meeting between CBers.

FANCY GAP Rt. 53 near Mt. Airy, West Virginia.

FAT LOAD overloaded; more weight than law allows.

FED INSPECTOR DOT or FCC.

FEED THE BEARS to get caught speeding or to pay a speeding fine.

FEED THE PONIES lose money at the horse races.

FENDER BENDER 1. vehicle collision 2. wreck.

FIFTY-DOLLAR LANE passing lane.

FINAL end of transmission.

FIND A CLEAN ONE switch to channel with less conversation.

FINGER WAVE an obscene gesture made with a finger.

FINGERS a channel-hopping CBer.

FIREWORKS police car lights.

FIRST SERGEANT wife.

FISH Plymouth Barracuda.

FIVE-BY-FIVE strong signal. (Also, KICKING OUT FIVE.)

FIVE-FINGER DISCOUNT stolen goods.

FIVE-FIVE 55 mph.

FIX police location report, location report of any kind.

FIX DAILY Ford Truck.

FLAG WAVER highway worker.

FLAG WAVER TAXI highway repair truck.

FLAPPER CB antenna.

FLASHLIGHT SHIRT western shirt with pearlsnap buttons worn by a truck stop commando.

FLATBED tractor trailer with flatbed.

FLATKEY 55 mph speed limit.

FLATKEYING jamming a channel by steady depression of mike switch, without talking.

FLIGHT MAN weigh-station worker.

FLIP OR FLIP-FLOP return trip, "See you on the flip flop."

FLIP-FLOPPING BEARS police reversing directions.

FLIPPER return trip.

FLOP IT turn around.

FLOWERS a good deed, after receiving a Smokey report you might say, "Thanks for the flowers."

FLUFF STUFF snow. Also Ivory Flakes.

FLYING KITES material falling off back of truck.

FLY IN THE SKY police helicopter. Also BEAR IN THE AIR.

FOG LIFTER interesting CBer.

FOLDING CAMERA a police car equipped with Vascar speed measuring equipment.

FOOT IN THE CARBURATOR going fast.

FOOT WARMER same as LINEAR AMPLIFIER.

FOR SURE that's right.

FORTY-FOOTER same as 18-wheeler.

FORTY-FOURS children, also kisses.

FORTY WEIGHT beer.

FOUR right, OK.

FOUR-BANGER four-speed transmission.

FOUR-BY-FOUR a Bronco, Blazer or Jeep with a 4′ x 4′ cargo area.

FOUR-LANE PARKING LOT crowded expressway.

FOUR LEGGED BEAST race horse.

FOUR LEGGED GO-GO DANCERS pigs.

FOUR ROGER message received. Also TEN ROGER, ROGER DODGER.

FOUR-TEN . 10-4 (emphatically); right on!

FOUR WHEELER a passenger car.

FOX CHARLIE CHARLIE Federal Communications Commission.

FOX HUNTING FCC looking for CBers breaking regulations.

FOXY LADY attractive woman.

FREE RIDE prostitute.

FREIGHT TRAIN tractor with two or three trailers.

FRIENDLY CANDY COMPANY The FCC, especially FCC headquarters in Washington.

FRISKIE CITY San Francisco, California.

FRONT DOOR lead rig in line of two or more trucks, see CONVOY.

FRUIT LINE a Freightliner truck.

Fugitive

FUGITIVE CBer operating on channel other than the favorite in the area.

FULL OF VITAMINS powerful, as in a big engine.

FUNNY BOOKS 1. pornography. 2. trucker's log book.

FUNNY BUNCH OF IDIOTS Federal Bureau of Investigation.

FUNNY BUNNY disguised police car.

FUNNY FARM retirement home, especially for old truckers.

GANDY DANCER road construction worker.

GAT gun. Also, BLOWER, BLOWGUN, POPPER, PEASHOOTER.

GATEWAY CITY St. Louis, Missouri.

GBY God Bless You.

GEARJAMMER truck driver.

GENERAL MESS OF CRAP General Motors Corporation truck.

GENERATING transmitting, broadcasting over CB.

GEORGIA OVERDRIVE neutral gear.

GET HORIZONTAL go to sleep, go to bed.

GET TRUCKING make some distance.

GET YOUR SIGNALS OFF have a good time with your CB.

GETTING OUT being heard by other CBers, also transmitting.

GIRLIE BEAR policewoman.

GIVE A SHOUT 1. answer back. 2. call a certain person.

GLITCH 1. an indefinable technical defect in CB equipment. 2. Also

a CB gremlin.
GLORY CARD CB license.
GLORY WAGON wildly decorated semi or truck.
GO same as DOWN AND OUT, "We go."
GO AHEAD answer back.
GO BACK talk again.
GO, BREAKER permission for a BREAKER to speak on the channel.
GO-GO GIRLS cows in a cattle truck.
GO JUICE or GO-GO JUICE gas, fuel, especially diesel fuel.
GOIN' HOME GEAR high gear.
GOLDEN ARCHWAY St. Louis, Missouri.
GOLDIE LOCKS traveling business woman.
GONE going off the air.
GONE TEN-SEVEN deceased. "He's gone ten-seven after his big accident."
GOOD BUDDY another CBer, but never a RUBBERBANDER.
GOOD LADY feminine equivalent of GOOD BUDDY.
GOOD NUMBERS best wishes, especially 73's and 88's.
GOODIED UP heavily accessorized.
GOODIES extra accessories for a CB.
GOOD SHOT road clear of police and obstructions.
GOON SQUAD channel hogs.
GOONEY BIRD machine that blocks sound with a pulsating tone.
GOT A COPY do you hear?
GOT HIS SHOES ON going full speed.
GOT MY EYEBALLS PEELED I'm looking.
GOT MY FOOT IN IT accelerating.
GO TO 100 head for rest-room stop.

Gooney Bird

GRANNY GEAR low gear.
GRASS the shoulder of the road, also the median divider.
GREASE SPOT a trucker who has met the Great Dispatcher in the Sky.
GRAVEL BUGGY dump truck.
GREAT BIG SPROCKET big engine.
GREEN CBer military police with CB.
GREEN HOUSE bank.
GREEN LIGHT road clear of police and obstructions.
GREEN MACHINE military truck.
G.S. great signal!
GREEN STAMPS money, dollars, especially those used for paying tickets: also the actual speeding tickets.
GREENS same as above.
GREEN STAMP COLLECTOR traffic court judge.
GREEN STAMP HIGHWAY toll road.
GREEN STAMP LANE passing lane.
GROUND CLOUDS fog.
GROWED UP TRUCKS tractor trailers.
GUARANTORY definitely, guaranteed.
GUITAR TOWN Nashville, Tennessee.
GUMPS stolen chickens.
GUN RUNNER police radar.
GUTTER BALLING bowling.
GUY fellow CBer. Also GOOD GUY, GOOD OLD GUY.
HAG FEAST a group of female CBers hogging a channel and chattering away. Also, KAFFEE KLATCH.
HAIRCUT PALACE bridge with low clearance.
HALL HAUL Hall's truck.

Hamster

HALLOWEEN MACHINE Cooper-Jarrett truck.
HAMBURGER HELPER same as LINEAR AMPLIFIER.
HAMMER accelerator.
HAMMER DOWN to accelerate. Also, HAMMER ON.
HAMMER OFF to slow down. Also HAMMER UP.
HAMSTER one who "hams" on CB, especially by attempting to "skip work."
HANDLE CBer's nickname.
HANG A RIGHT (LEFT) turn right (or left).
HANG A UEY make a U-turn.
HANG IT IN YOUR EAR take your signals and. . . .
HANG OUT monitor a specific channel.
HANGER garage.
HAPPY NUMBERS an S-meter reading, especially, a five or maximum output reading.
HARD TO PULL OUT difficult to understand.
HARVEY WALL BANGER reckless driver.
HASH AND TRASH background noise, an unclear signal.
HAVE A SAFE ONE AND A SOUND ONE drive safely.
HEDGEHOPPER pick up truck.
HENCH MEN a group of CBers.
HIDING IN THE BUSHES, SITTING UNDER THE LEAVES police hiding, especially on the side of road.
HI-BALL HEAVEN truck stop.
HIGH GEAR linear amplifier in use.
HILLBILLY CHROME home-done paint job.
HILLBILLY OPERA HOUSE CB radio.
HIND END last CB vehicle in a line of two or more, also DONKEY.
HIT THE HAY go to sleep.
HITCH-HIKER a vehicle without CB following a CBer.
HO CHI MINH TRAIL Route 209, through the Pennsylvania coal country.
HOG COUNTRY Arkansas.
HOLE IN THE WALL tunnel. Also, Wheeling, West Virginia.
HOLDING ON TO YOUR MUD FLAPS driving right behind another vehicle.
HOLLER 1. call. 2. invitation to transmit.
HOLSTEIN a black and white state trooper's car.
HOME CHANNEL a channel chosen by two or more people and used regularly by prearrangement.
HOME PORT residence location. See HOME 10-20.
HOME 10-20 same as above.
HONEY BEAR policewoman.
HONEY BUCKET mobile latrine.
HONEY BUCKET STOP roadside restroom.
HONEY WAGON beer truck.
HOOKER HAVEN whore house. Also, certain truck stops, eg., Naked City, Indiana.
HOO-HOOER airhorn.
HOO HOONER left-lane hog.
HOPPINS stolen vegetables.
HORSE 1. Ford Mustang or Dodge Colt 2. tractor trailer.
HOT LANTA Atlanta, Georgia. Also, HOT TOWN.
HOT PANTS 1. smoke or fire. 2. the driver's seat.
HOT STUFF coffee.
HOT WATER CITY Hot Springs, Arkansas.
HOT WIRE same as LINEAR AMPLIFIER.
HOW ABOUT (NAME) ONE TIME? call for a specific CBer.

HOW ABOUT YOUR VOCAL CORDS? "Is your CB working?"
HOW AM I HITTING YOU? "How well do you receive my signal?"
HOW DO YOU READ ME? "What is the S-meter reading on my signal?"
HOW TALL ARE YOU? "What is the height of your truck?"
H. TOWN Hopkinsville, Kentucky.
HUNDRED MILE MUD extremely strong black coffee (flavor lasts 100 miles).
HYDROPLANE truck skidding on puddles or wet pavement.
ICE BOX refrigerator truck.
I'M THROUGH "I'm finished transmitting."
IN A SHORT SHORT soon.
INDIAN neighbor who has interference on his TV from your CB.

Indian

IN THE BUSHES off the road.
IN THE GRASS on the median divider or shoulder.
INVITATION police summons.
IT'S ALL CLEAR "No bears around."
JACK IT UP accelerate.
JACK RABBIT 1. speeding police car. 2. speeder of any kind.
JAKE BREAKS truck brakes.
JAMBOREE large gathering of CBers, often including camp outs, entertainment, prizes, and equipment displays, usually sponsored by CB club or radio station.
J. TOWN Jackson, Tennessee.
J. TRAIL CB jamboree season, usually February to November.
JAVA coffee.
JAW JACKING talking over CB, particularly for more than the legal five minutes.
JIMMY GMC truck.
JUICE MAN trucker disc jockey, also BIG JUICER.
JUNK BUZZARD a bum among bums.
JUNKYARD place of employment.
KEEP 'EM BETWEEN THE DITCHES "Drive safely."
KEEP THE ROLLING SIDE DOWN AND THE SHINY SIDE UP, also KEEP THE SHINY SIDE UP AND THE GREASY SIDE DOWN "Drive safely, don't tip over."
KEEP THE WHEELS SPINNING "Drive safely."
KEEP YOUR NOSE BETWEEN THE DITCHES AND SMOKEY OUT OF YOUR BRITCHES "Drive safely and watch out for speed traps."
KENOSHA CADILLAC any car manufactured by American Motors (AMC).
KEYBOARD controls on CB set.

Keyboard

KEYING THE MIKE activating the mike without speaking, channel jamming.
KICKER same as LINEAR AMPLIFIER.
KICKING transmitting.
KIDDIE CAR a school bus.
KIDDIE COP a CBer who likes to play policeman.
KIDNEY-BUSTER a truck.
KITTY WHOMPER a Kenworth tractor. Also, K. WHOMPER, KANTWORK.
KNOCK THE SLACK OUT accelerate.
KNUCKLE BUSTER a fight.
KODAK police radar.
KODIAK a state trooper, also KOJAK.
KOJAK WITH A KODAK police with radar.
KOOL AID beer.
K. TOWN Knoxville, Tennessee.
LACE CURTAIN fancy hotel or motel.
LADY BEAR policewoman.

LADY BREAKER female CB operator, asking permission to use a channel.

LAND LINE telephone.

LAND OF DISNEY Disneyland, California. Also, California.

LAND OF WONDERFUL "road is clear of Smokies and obstructions."

LATCH-ON vehicle without CB following one with CB.

LATRINE LIPS one who uses obscene language on CB.

LAY AN EYE ON take a look at, "lay an eye on them seat covers."

LAY 'EM DOWN WHERE YOU ARE stop, trouble or police ahead.

LAY IT OVER stand by.

LEFT SHOULDER the opposite direction. Also, OVER YOUR SHOULDER.

LEGAL BEAGLE one who uses his legal call letters and abides by FCC rules.

LET IT GO drive at desired speed.

LET IT ROLL accelerate.

LET THE CHANNEL ROLL let others break in and use the channel.

LET THE HAMMER DOWN drive full speed, road is clear of Smokies and obstructions.

LETTUCE money.

LET YOUR FLAPS DOWN slow down.

LID inept operator.

LIGHTS GREEN, BRING ON THE MACHINE road is clear of police and obstructions, drive on.

LIL' OL' MODULATOR CB set.

LIMP LINE loose rigging.

LINEAR AMPLIFIER illegal amplifier which boosts CB transmitter output.

LISTENING IN monitoring.

LIT CANDLES police-car lights turned on.

Little Bears

LITTLE BEARS local police.

LITTLE BIT 1. prostitute 2. sexual adventure.

LITTLE BLACK BOX same as LINEAR AMPLIFIER.

LITTLE FOOT WARMER same as LINEAR AMPLIFIER.

LITTLE MAMA short antenna.

LOAD OF ROCKS truck hauling bricks.

LOAD OF STICKS truck hauling timber.

LOAD OF VW RADIATORS truck traveling empty.

LOCAL BEARS local police.

LOCAL BOY local police.

LOCAL YOKEL local police.

LOG SOME Z'S get some sleep.

LOLLIPOP desk mike.

LONG DISTANCE TELEPHONE IS RINGING call for a specific CBer.

Long Nose

LONG NOSE diesel with engine in front of cab, not under it, (cf. COE).

LOOKING GOOD putting out a strong signal.

LOOSE BOARD WALK bad road.

LUMBER BUMPER log-carrying truck.

LUNCH BOX CB rig.

MAGNOLIA STATE Mississippi.

MAIL CB conversation.

MAKE A TRIP switch channels.

MAKING THE TRIP transmitting powerfully, being heard.

Mama

MAMA wife, (*never* mother).

MAMA BEAR police woman.

MAMA SMOKEY female state trooper.

MANIAC mechanic, especially truck mechanic. MINI MANIAC is an auto mechanic.

MAN IN BLUE any policeman.

MAN IN SLICKER fireman. Also, MAN IN RED.

MAN IN WHITE doctor.

MAN WITH A GUN police with radar.

MARDI GRAS TOWN New Orleans, Louisiana.

MARK-L OR MARK-EL-MAN man working for the Markel Insurance Company, who checks speeding trucks, and makes reports to the company.

MARKER milepost along interstate, used to determine and report driver's exact location.

MASHING THE MIKE same as KEYING THE MICROPHONE.

MAYDAY distress call. (10-34.)

MEAT WAGON ambulance.

MERCY wow!—critically over-used CB expletive.

MERCY SAKES in all circumstances, means simply "mercy sakes."

MEXICAN OVERDRIVE neutral gear, used going downhill, freewheeling.

MICKEY MITCHELL local police.

MICKEY MOUSE METRO (ON A TRICYCLE) local police (on a three-wheel motorcycle).

MIDNIGHT OVERDRIVE same as "MEXICAN OVERDRIVE."

MIDNIGHT SHOPPER thief.

MIK-E-NIK mechanic.

MILE HIGH CITY Denver, Colorado.

MILE POST MARKER along road used to determine and report exact location.

Mini Skirt

MINI SKIRT woman, girl.

MINI STATE Rhode Island.

MIX MASTER highway clover leaf.

MIXING BOWL 1. highway clover leaf. 2. converging roads.

MOBILE 1. CB transceiver designed for use in car, truck, boat, plane, car, etc. 2. a CBer in a vehicle 3. vehicle with CB.

MOBILE PARKING LOT auto carrier.

MOBILING going for a ride. Also, moving or on the road.

MOCCASINS same as LINEAR AMPLIFIER.

MODULATE talk.

MODULATION conversation.

MOLLIES uppers or sleep-retarding pills.

MONFORT LANE passing lane, named after Monfort trucks.

MONITOR to listen, but not participate in CB conversation; to standby, waiting for a call.

MONKEY BLOOD fuel, especially gasoline.

Monkey Town

MONKEY TOWN Montgomery, Alabama.

MONKEY WARD Montgomery Ward store.

MONSTER LANE inside passing lane.

MOTHBALL annual CB convention. Also JAMBOREE.

MOTION-LOTION fuel, either diesel or gasoline.

MOTOR CITY Detroit, Michigan.

MOTOR IS TOTERING driving at desired speed.

MOTOR MOUTH constant talker.

MOTORING traveling on without regard to police and/or road conditions.

MOUNTIES IN THE SKY police in helicopter. Also, BEARS, BIRDS, OR MOUNTIES IN THE TREE.

MOVABLE PARKING LOT automobile transport.

MOVING FOREST log-carrying truck.

M-TWENTY meeting place.

MUD coffee. Also JAVA, BROWN BEER, et al.

MUFF woman, girl.

MUSCLE BUSTER air-cushioned (Bostrum) truck driver's seat.

MUSIC CITY Nashville, Tennessee. Also GUITAR CITY, NASTYVILLE.

MUSKRAT child.

NAP TRAP rest area or motel.

NASTYVILLE Nashville, Tennessee.

NATURE STOP rest-room stop.

NEGATIVE CONTACT station being called but not responding. Also, "Nobody answered my call."

NEGATIVE COPY no answer.

NEGATORY no. Opposite of POSITORY.

NIGHT CRAWLERS unmarked police cars, operating after dark.

NINE-TO-FIVER truck driver who only makes short runs, or who works a delivery driver.

NOBODY KNOWS WHERE THE TEDDY BEAR GOES "Smokies are criss-crossing the highway."

NUMBERS, THE NUMBERS 7s & 3s; salutations; also LAY THE NUMBERS, THE BIG NUMBERS.

OIL BURNER car with smoking exhaust; car in need of ring-job.

OM (OLD MAN) any CBer.

ON A (CITY NAME) TURN return trip from named city. "I'm on an Atlanta turn."

ON STAND BY listening, but not transmitting; also on the by.

ON THE PAY going at legal speed.

ON THE PEG Same as above.

ON THE SIDE 1. parked. 2. monitoring a channel and joining in once in a while.

ON THE SIXTY going 60 mph. Also, DOUBLE NICKELS PLUS FIVE.

ONE the original. "This is the One Rubber Duck."

ONE-EYED MONSTER a TV set.

One Eyed Monster

ONE FOOT ON THE FLOOR, ONE HANGING OUT THE DOOR, AND SHE JUST WON'T DO MORE "Mercy sakes, I'm flying!"

ONE HIDING IN THE GRASS police in the median.

ONE TIME a brief CB contact.

OPEN SEASON police are everywhere.

OTHER HALF a CBer's wife. Also BETTER HALF.

OTHER RADIO special radio to receive police transmissions. Also, a SCANNER.

OUT through transmitting.

OUT OF IT signing off. "I'm out of it."

OUTDOOR TV drive-in theatre. Naturally, INDOOR TV is a movie house.

OVER through transmitting. "I'm over. Your turn to transmit."

OVER YOUR SHOULDER on the road directly behind you.

OVERBANDING broadcasting on an illegal frequency.

OVERMODULATNG incoming voice is muffled or whistling, often caused when a preamp microphone is turned too high.

OW (OLD WOMAN) The FCC mobile monitoring unit is nearby.

PA public address system.

PACK IT UP finish.

PADIDDLE car with one headlight.

PAINTING THE TOWN PINK issuing FCC violation tickets.

PAIR OF SEVENS no contact or answer.

PANIC IN THE STREETS the FCC is monitoring area.

PANTY STRETCHER fat lady.

PAPA BEAR state trooper with CB.

Pair of Sevens

PAPER HANGER police giving speeding tickets.
PAPER WORK speeding ticket.
PART YOUR HAIR be on your best behavior, police ahead.
PASS THE NUMBERS TO YOU to send out 73's and 88's
PATCH a city or town.
PAY HOLE GEAR high gear, same as GOING HOME GEAR.
PEAKED UP tuned up.
PEAK POWER maximum wattage.
PEANUT BUTTER IN HIS EARS not listening.
PEDAL to coast.
PEDAL TO THE METAL accelerate.
PEDAL A LITTLE SLOWER slow down.
PEDAL TO THE METAL accelerate. "Put the pedal to the metal and eat up those white lines."

Pedal to the Metal

PEDALING IN THE MIDDLE 1. straddling two lanes. 2. driving in the middle lane of a three-lane highway.
PEEPING IN monitoring.
PEELING OFF 1. getting off the expressway. 2. making a turn.
PENMAN a CBer to be who has filed for his FCC license.
PENNSLOMANIA Pennsylvania, where the speed limit was 55 mph even before the fuel shortage.
PEPPERMILL cinder-spreading truck in winter.
PERSUADER same as LINEAR AMPLIFIER.
PETE Peterbilt truck.
PETER COTTONTAIL male hitchhiker "with ears."
PETER RABBIT police of any kind.

Peter Rabbit

PETRO REFINERY truck hauling gas or oil
PF FLYERS truck wheels.
PICK-EM-UP PUP pickup truck.
PICKING PEAS police using radar. "Them bears is PICKING PEAS."
PICTURE BOX same as PICTURE TAKING MACHINE.
PICTURE (TAKING) MACHINE police radar unit.
PIECE OF PAPER speeding ticket.
PIGEON vehicle stopped by police.
PIGGY BACK a trailer attached to a car.
PIGGY BANK toll booth.
PIGGYBANK STATE Maryland.
PIGS police.
PIMPLE temporarily mounted flashing light, used on unmarked police cars, when in "hot pursuit."
PINK PANTHER unmarked police car; also police with CB.
PINK QSL CARD warning ticket.
PIPE LINE a specific channel.
PIT STOP fuel stop.
PLAIN BROWN CONVOY National Guard vehicles.
PLAIN BROWN (BLACK, WHITE) WRAPPER unmarked police car of any specific color.
PLAY DEAD standby.
PLAYING BASKETBALL monitoring a channel.
PLAYING TAG four-wheelers riding in middle of truckers' convoy.
PLOWING UNDER returning home with empty load.
POLACK KIDS cattle.
POLACK SCHOOL BUS cattle truck.
POLE CAT black and white highway patrol car.
POLAROID radar.

Porky Bear

POOR DEVIL newlywed. Also, SKINNED CAT.

PORCH LIGHTS spot lights or fog lights.

PORKY BEAR any police.

PORTABLE BARNYARD livestock truck.

PORTABLE CHICKEN COOP mobile weighing station.

PORTABLE FLOOR flatbed tractor trailer.

PORTABLE GAS STATION tank truck carrying gasoline.

PORTABLE PARKING LOT auto carrier.

PORTRAIT PAINTER police radar.

PORTABLE ROAD BLOCK McClean truck.

PORTABLE WAITING ROOM Greyhound bus. Also CARELESS CATTLE CAR.

POSITIVE yes. affirmative.

POST milepost. MARKER.

POSTHOLES (LOAD OF) empty flat bed. Also empty load.

POT HOLE STATE West Virginia.

POTTY MOUTH person who uses obscene language on CB.

POUND METER S-meter; see Technical Glossary.

POUNDS meter reading in "S" units.

POUR COFFEE ON YOU to buy a good buddy a cup of coffee.

POWER CITY Pittsburgh. Also STEEL TOWN.

PREGNANT ROLLER SKATE Volkswagen Beetle.

PRESS SOME SHEETS get some sleep.

PRESSURE COOKER sportscar.

PROFESSIONAL 1. trucker. 2. prostitute.

PTT push-to-talk, refers to mike button.

PULL IN FOR A SHORT make a rest-room stop.

PULL THE BIG ONE signoff, also PULL THE PLUG.

PULLING 'EM DOWN police pulling car off to side of road.

PULLING YOU OUT reading you fine despite interference.

PULL YOUR HAMMER BACK slow down.

PUMPKIN flat tire.

PUNCHED TICKET revoked FCC license.

Puppy Dog

PUPPY DOG Mack truck. Also BULLDOG.

PUSH AND WONDER brakes.

PUSHING A RIG 1. driving a truck. 2. operating a CB.

PUSH O LINE fuel.

PUSH WATER fuel.

PUT AN EYEBALL ON look at.

PUT IT ON THE FLOOR AND LOOK FOR SOME MORE to accelerate.

PUT THE GOOD NUMBERS ON YOU best wishes, 73's and 88's.

PUT THE HAMMER DOWN accelerate.

PUT YOUR BOOTS ON "Turn on your linear amplifier."

PUT YOUR FOOT ON THE FLOOR AND LET THE MOTOR TOTER accelerate.

PUT YOURSELF UP HERE "Move up the road; it's clear of police and obstructions."

Q-BIRD an intermittant tone generator.

QRM interference.

QSL CARD postcard bearing call sign of a CB station mailed to verify communication or exchanged at coffee break or club meeting.

QUEEN CITY Cincinnati, Ohio.

QUICK TRIP AROUND THE HORN scanning all the CB channels.

QUIZ breath test to determine how much alcohol has been consumed.

RADAR ALLEY Interstate 90, Ohio.

R&R rest and recuperation.

RATCHET JAW CBer who talks too much.

RATCHET JAWING talking; idle talking; talking too long.

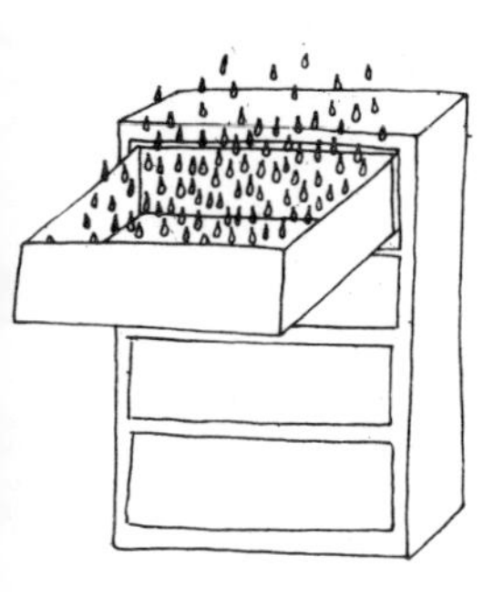
Rain Locker

RADIO CHECK a report on the quality of transmission.
RAIN LOCKER shower room.
RAKE THE LEAVES last CB vehicle in convoy who looks for police from the rear, see also: SHAKE THE TREES.
RALLY intermediate-size CB gathering.
RAT RACE heavy traffic.
READ to understand or receive a transmission.
READ hear, copy.
READ THE MAIL to monitor a channel.
REBOUND return trip.
RED WHEEL spinning red light on police car.
REEFER refrigerator trailer.
RELOCATION CONSULTANTS moving vans.
REPAIR DAILY Ford Truck.
REST 'EM UP rest area.
RIBBON highway. Also, any road.
RIDER vehicle without CB, following one with CB.
RIG 1. truck 2. CB transceiver.
RINGING YOUR BELL "Someone's calling you."
RIP STRIP Interstate highway or freeway.
RIVER CITY Memphis, Tennessee. AlsoPaducah, Kentucky.
ROAD JOCKEY driver of tractor trailer.
ROAD TAR coffee.
ROCK slang for crystal, the tuning device set to allow CB transceiver to receive specific channels.
ROCK CITY Little Rock, Arkansas.
ROCKING CHAIR CB vehicles in the middle of a convoy.
RODADIO radio.
ROGER yes, okay.
ROGER DODGER same as ROGER, also "Roger D."
ROGER RAMJET driver of car going well over speed limit.
ROGER ROLLER SKATE any car doing 20 miles or more over the speed limit.
ROLLERSKATE any small or foreign car.
ROLLING moving, especially, on the road.
ROLLING BEARS police on the move.
ROLLING REFINERY truck hauling gas or oil.
ROLLING ROAD BLOCK vehicle going under speed limit and holding up other traffic.
ROVER highway police on the move.
RUBBERBANDER 1. a new CBer (who doesn't know the CB language). 2. any deadhead, in general.
RUBBER CITY Akron, Ohio.
RUBBER NECKERS people who look out of car or truck windows, especially at an accident.
RUG RATS children.
RUMBLE SEAT vehicle without a CB, following one with CB.
RUNNER a police chase car with radar.
RUNNING SHOTGUN driving partner. Also, SITTING SHOTGUN.
RUNNING TOGETHER CBers staying in contact on the highway.
RUN OUT THE FRONT END FRONT DOOR in CONVOY pulls too far ahead and out of CB range.
SADDLE same as ROCKING CHAIR.
SAFE TRUCKIN' good trip.
SAFER SHAFFER Shaffer truck.
SAIL BOAT FUEL wind.

Saddle

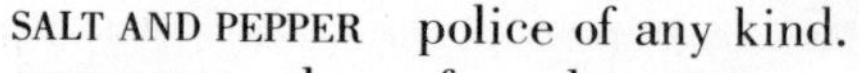

SALT AND PEPPER police of any kind.
SALT MINE place of employment.
SALT SHAKER salt spreading truck in winter.
S&H GREEN STAMPS money.
SANDBAGGING monitoring, but not transmitting.
SAN QUENTIN JAIL BAIT underage female, often a hitch-hiker.
SAVAGES CBers who hog a channel.
SAY WHAT repeat, "What did you say?"
SCALE HOUSE truck weigh station.
SCALE HOUSE IS ALL RIGHT weigh station is closed.
SCATTER STICK omnidirectional CB antenna.
SCHOOL TWENTY location of school or college.
SCOFFLAW FCC rule violator.
SEAT COVER woman in an automobile.
SET OF DIALS CB controls.
SET OF DOUBLES tractor-trailer truck.
SEVEN THREE a combination of the ten code meaning signing off ("Ten Three" means stop transmitting and "Ten Seven" means out of service, leaving air).

Seat Cover

SEVENTY-THREE(S) best wishes.
SHACK railroad conductor.
SHAMUS cop.
SHAKE THE BUSHES the lead CB vehicle in a convoy who checks for police and obstructions. Also SHAKE THE TREES.
SHAKEY CITY (TOWN) Los Angeles, California.
SHAKEYSIDE West Coast.
SHAKING IT moving fast.
SHAKING THE WINDOWS signal is coming in loud and clear.
SHANTY SHAKER mobile-home driver.
SHARK TOWN Long Island.
SHOAT AND GOAT CONDUCTOR trucker who carries livestock.
SHOES Same as LINEAR AMPLIFIER.
SHOOT AN EYE BALL to see, look at.
SHOOTING SKIP talking beyond the 150 mile legal distance. Also WORKING SKIP.
SHOPPING AROUND checking the channels for other CBers to talk to.
SHORT SHORT soon.
SHORT SHOUT brief CB conversation.
SHOT GUN 1. police radar device that looks like rifle. 2. seat adjacent to the driver, as in riding SHOT GUN.
SHOT OUR FIVE to go beyond the FCC five minute limit on continuous conversation.

Shot Gun

SHOUT call another CBer.
SHOVELLING COAL accelerating.
SHOW-OFF LANE passing lane.
SHY TOWN Chicago.
SIDE DOOR passing lane.
SIDEWINDER single side band user.
SIN CITY Cincinnati, Ohio, Also Las Vegas, Nevada.
SINGLE SIDEBANDER •CBer who operates on a side band, (see Technical Glossary).
SINGING WAFFLES radial tires.
SITTING IN THE SADDLE middle CB vehicle in a line of three or more, see ROCKING CHAIR.
SITTING ON THE BACK PORCH parked near tunnel or underpass.
SITTING ON THE FRONT PORCH located on bridge or overpass.

SITTING ON THE SWING 1. located in truck stop or parking lot. 2. driving in the rocking chair.

SIX-LEGGED SMOKEY policeman on horseback.

SIX WHEELER small truck or car pulling trailer.

Silver Bugle

SIGNALS OFF similar "to getting your rocks off," having a good time. "I got my SIGNALS OFF at the jamboree."

SILVER BUGLE air horn.

SKATING RINK slippery road.

SKIP to contact far away stations by bouncing signals off the ionosphere.

SKIP LAND radio stations hundreds of miles away.

SKIPPERS CBers who talk on long-range signal.

SKIP SHOOTER unlicensed CBer.

SKIP TALK long range conversation, created by bouncing signals off ionosphere.

SKY BEAR police helicopter. Also SKY MOUNTY.

SLAMMER jail.

SLICK TOP an unmarked police car with no light on top.

SLIDER an illegal CB device which permits transmission on unauthorized CB channels.

SLIP THE SEAT to change vehicles.

SLOP bad fuel, watered-down gasoline.

SLOPPY JOES state troopers.

S-METER signal strength indicator, see Technical Glossary.

SMILE AND COMB YOUR HAIR radar ahead, slow down.

SMOKE 'EM OUT lure out hidden police cars by exceeding speed limit only slightly.

SMOKE 'EM UP BEAR police.

SMOKE ON, BROTHER accelerate.

SMOKE (SMOKEY) REPORT police location report.

SMOKE SCREEN police radar.

SMOKE SOME DOPE accelerate.

SMOKER same as SMOKEY THE BEAR.

SMOKEY (THE BEAR) police of any kind.

SMOKEY BEAVER policewoman.

SMOKEY BOOK a directory showing police communication channels.

SMOKEY DOZING police in parked car.

SMOKEY GRAZING GRASS police in median strip.

SMOKEY ON RUBBER police moving.

SMOKEY'S THICK police are everywhere.

SMOKEY TWO WHEELER motorcycle cop.

SMOKEY WITH A CAMERA police with radar.

SMOKEY WITH EARS police with CB.

SNAFU foul up.

SNAKE DEN fire station.

SNAKE SNIFFER person looking for trouble.

SNEAKY SNAKE hidden patrol car. Also police with CB.

SNOOPER SCOPE an illegally high antenna, over the maximum 60-foot limit.

SNOW BUNNY skier.

SNUFFY SMITH Smith Transfer Company driver.

SOCKS same as LINEAR AMPLIFIER.

SOMEONE SPILLED HONEY ON THE ROAD police are everywhere.

SONNET a CBer who advertises products on the air.

SOUPED UP modified rig with illegally high CB power.

SPAGHETTI BOWL highway interchange. Also MIXMASTER.

Sneaky Snake

SPARKIE electrician.
SPLASH OVER BLEEDING.
SPLATTER Same as BLEEDING.
SPLIT BEAVER stripper or go-go dancer.
SPLIT THE SHEET to get a divorce or separation.
SPLIT YOUR SIDES transmit on single-side band.
SPOKE TO US answered back.
SPORT CITY Shreveport, Louisiana.
SPREADING THE GREENS police are passing out speeding tickets.
SPY IN THE SKY police helicopter.
SQUARE WHEEL DUCK TRUCK camper or recreational vehicle so overloaded with accessories and equipment it moves as if it has square wheels.
SQUAWK BOX CB radio.
STACK THEM EIGHTS best regards.
STAGE STOP truck stop.
STARVE THE BEARS don't get a ticket.
STATE BEAR state trooper.
STEPPED ALL OVER YOU interrupted your transmission.
STEPPED ON THE BEAR'S TOES broke the law, speed limit.
STEPPING moving.
STINGER antenna, especially a center or top-loaded model.
STOP TO GET GROCERIES stop and eat.
STRAIGHT SHOT road clear of police and obstructions.
STREAKING going at full speed. Also moving.
STRETCHED IT OUT picked up speed.
STROLLER CBer with a walkie-talkie.
STRUGGLE attempt to BREAK into a channel.
STUFFY congested.
SUCKER a CB rig on the service bench.
SUDS beer.
SUGAR BEAR policewoman.
SUGAR FOOT speeder who never gets caught. Also FEATHERFOOT.
SUICIDE CARGO dangerous cargo.
SUICIDE JOCKEY trucker carrying explosives.
SUNBEAM a CBer who livens the channel with his humor, wit, and great stories. "Oh, for a SUNBEAM, for a SUNBEAM bright!"

Super Chicken

SUNOCO SPECIAL New York State police car.
SUPER CHICKEN Yellow Freight System truck.
SUPERDOME CITY New Orleans, Louisiana.
SUPER SKATE sports car.
SUPER SLAB expressway, interstate highway.
SUPERSTRUCTURE bridge.
SUPPOSITORY negatory.
SURF to have a good time on the air waves.
SURFER experienced CBer.
SWEEPING LEAVES last truck in a CONVOY watching out for police from the rear. Also RAKING THE LEAVES.
SWEET THING female CBer.
TAGS license plates.
TAKE IT BACK your turn to talk.
TAKING PICTURES police using radar.
TAKING PICTURES EACH WAY police using radar in both directions.
TALKING THROUGH YOUR HAT BLEEDING onto an adjacent channel.
TANKER trailer carrying liquid.
TATTLETALE police in sky.

Tanker

TAXI police car.

TEAR JERKER CBer who sings the blues, about equipment, rubber-banders, etc.

TENDERFOOT limited-channel CB set.

TENNIS SHOES truck tires.

TEN-ONE HUNDRED (10-100) restroom stop.

TEN-ONE THOUSAND (10-1000) FCC man.

TEN ROGER message received.

TEN-TEN, TILL WE DO IT AGAIN "Signing off and see you soon."

TEN-TWO THOUSAND (10-2000) dope dealer.

TEN, BYE BYE good-bye.

TEN FOUR HUNDRED drop dead.

TEN POUNDER excellent radio.

TENSE congested traffic.

TEXAS STRAWBERRIES shelled corn.

THE DIRTY SIDE New York City, New Jersey.

THERMOS BOTTLE tank truck.

THIRTY TWELVE (30-12) ten four, three times.

THIRTY WEIGHT coffee.

THREAD 1. wire in CB set. 2. cable connecting CB set and antenna.

THREES AND EIGHTS (73s AND 88s) best regards, hugs and kisses!

THREES ON YOU best regards!

THROW A FIT use a linear amplifier.

THROWING transmitting.

THROWING A CARRIER Same as keying the MICROPHONE.

THROWING NINE POUNDS AT ME strong signal that reads "9" on the S-meter.

Ticks

THUNDER CHICKEN Ford Thunderbird.

TICKET PUNCHER FCC informer who turns in violators.

TICKS minutes, each tick is one minute.

TIGER IN A TANK Same as linear AMPLIFIER..

TIGHTEN UP ON THE RUBBERBAND accelerate.

TIGHTEN YOUR SEAT DOWN, WE'RE RUNNING HEAVY "Brace yourselves, we're accelerating."

TIAJUANA TAXI police, ambulance or wrecker especially when emergency lights are flashing.

Tin Can

TIN CAN CB rig.

TINSEL TOWN Hollywood, California.

TOE NAILS ARE SCRATCHING accelerating. Also TOE NAILS IN THE RADIATOR. Also TOE NAILS ON THE FRONT BUMPER.

TOILET MOUTH CBer who uses dirty language on the air.

TOOLED-UP a souped-up rig.

TOP TWENTY National CB Jamboree which is held for three days in a different city each year.

TOPSIDE roof of your house or top of your car.

TRADING STAMPS money.

TRAFFIC CB conversation.

TRAIN STATION 1. small town traffic court that issues a lot of fines. 2. kangeroo court.

TRAINING WHEELS learner's permit.

TRAMPOLINE bed.

TRANSCEIVER combination radio transmitter and receiver.

TRANSPORTER any truck.

TRICK BABE prostitute.

TRICKY DICK'S San Clemente, California.

TRIP clear CB signal heard from a distance. Also, any transmission.

TRUCK 'EM EASY "Drive safely."
TRUCK 'EM UP STOP truck stop.
TRUCKIN' GUY fellow truck driver, a good buddy.
TRUCKIN' TEENAGER teenaged hitchhiker.
TRUCK STOP COMMANDO duded-up trucker with fancy clothes.
TRUCKIN' TEENAGER teenaged hitchhiker.
TRUCKOLOGIST same as truck stop COMMANDO.
TUCK IT IN move into right lane.
TUNED UP a radio putting out more than 4 watts.
TURKEY dumb CBer.
TURKEY AREA rest area.
TURN AROUND return trip.
TURN OVER stop.
TURN TWENTY location of next exit or turn.
TURNING MY HOUSE AROUND rotating antenna for clearer reception.
T V I TV interference from CB transmission.
TWELVES company president.
TWENTY location.
TWENTY BUSHEL twenty ton payload.
TWIN HUSKIES antennas mounted on both rearview truck mirrors on a cab.
TWIN MAMAS dual antennas.
TWIN PIPES dual exhaust.
TWISTER highway interchange.
TWO MILES OF DITCHES FOR EVERY MILE OF ROAD "Keep your rig in the middle of the road and drive safely."
TWO WHEELER motorcycle or motorbike.
TWO WHEEL WINDMILL bicycle.
ULCER congested traffic.
UNCLE CHARLEY the FCC.
UNDRESSED 1. not using linear amplifier. 2. unmarked law enforcement vehicle.
UNGOWA BWANA "Howdy, Mister!"
UP ONE move up one channel.
USE THE JAKE "Slow 'er down."
VAN tractor-trailer truck.
VOICE CHECK radio check.
VOLKSWAGEN SPOTTER small, convex truck mirror.
WALKED ON signal covered up by a stronger signal from someone else.
WALKIN' T walkie talkie.
WALKIE TALKIE 1. hand-held transceiver. 2. broken-down truck with CB rig.
WALKING IN HERE, BLOWING SMOKE transmitting powerfully.
WALKING ON YOU covering up your transmission.
WALKING TALL a strong, clear reception.
WALKING THE DOG clear reception, transmitting.
WALL PAPER Same as QSL card.
WALL TO WALL 1. a superpowerful CB signal. 2. everywhere.
WALL TO WALL AND TREE TOP TALL receiving you loud and clear; also blowing smoke, loud and clear.
WALL TO WALL AND TEN FEET TALL clear reception of signal.
WALL TO WALL BEARS police all over the place.
WALLACE LANE middle lane in three-lane highway.
WARDEN 1. wife. 2. FCC.
WATCH THE PAVEMENT drive safely.
WATCH YOUR DONKEY watch the rear for police.

Turkey

Undressed

Week-end Warrior

WATER HOLE truck stop.
WATERGATE CITY Washington, D.C.
WAVE MAKER water bed.
WAY IS BUENO road is free of police and obstructions.
WE DOWN, WE GONE, BYE-BYE finished talking.
WE GONE GOOD-BYE just listening.
WEAR YOUR BUMPER OUT to follow too closely.
WEARING SOCKS using a linear amplifier.
WEAR OUT finish talking, "I'm WEAR OUT."
WEEK-END WARRIORS National Guard.
WEIGHT WATCHER weigh-station worker.
WELFARE STATION CB purchased with welfare payment.
WE'LL BE CLEAN ON YOURS signing off.
WE'RE BACKING 'EM UP NOW signing off, slowing down.
WE'RE CLEAR signing off. Also road is clear of police and obstructions.
WE'RE DOWN signing off. Also, WE'RE OUT.
WE'RE DOWN AND ON THE SIDE through transmitting, but still listening.
WE'RE LISTENING answer back. Also WE'RE LOOKING.
WE'RE TRYING attempting to make distant or difficult CB contact.
WEST COAST MIRRORS mirrors on both sides of cab.
WEST COAST TURNAROUNDS amphetamines that will keep you awake from coast to coast.
WHAMMING uttering obscenities while holding the mike open.
WHAT AM I PUTTING ON YOU? request for S-meter reading.
WHAT ARE YOU PUSHING? 1. What are you driving? 2. What kind of CB are you using? 3. What kind of antenna are you using?
WHATEVERS state troopers.
WHAT'S YOUR TWENTY? "Where are you?"
WHEEL BARRON dump truck.
WHEELS mobile CB.

Wheels

WHEN DID YOU GET IN THIS BUSINESS How long have you had a CB?
WHERE DO YOU GET YOUR GREEN STAMPS? Where do you work?
WHIP long antenna.
WHO DO YOU PULL FOR? Who do you work for?
WHOMPING jamming or interrupting transmission.
WIERDY same as RUBBERBANDER.
WILLY WEAVER drunk driver.
WIMP man with little courage or color, a RUBBERBANDER.
WIND JAMMER long-winded CBer.
WINDOW WASHER rainstorm.
WINDY CITY Chicago, Illinois.
WOOD BUTCHER carpenter.
WOOLY BEAR woman. Also WOOLY-WOOLY.
WORK TWENTY place of employment.
WORKING MAN truck driver.

Windjammer

WORKING THE MOONLIGHT EXPRESS running back roads at night to avoid weigh stations.
WORKING WOMAN prostitute.
WRAPPER the color of a vehicle, especially a police car.
WRINKLE uneven transmission or reception.
W.T. walkie talkie.
X-RAY MACHINE police radar.
XYL wife.
XYM husband.
YAK SHACK radio room.
YAP talk.

Yo Yo

YELLOWSTONE PARK 1. a gathering of police. 2. a road block or radar with several chase cars.
Y L young lady.
YO yes.
YO YO vehicle varying speed.
YOU GOT IT go ahead, I hear you.
ZAPPING transmitting powerfully, possibly so powerful as to cause damage to receiving unit.
ZOO bear headquarters.

TRADITIONAL TEN CODE

This code, still used by older CBers, is no longer recommended because it has many more entries than are needed. Note conflicts—for example, 10-24 versus 10-36 (time check)—with the new revised and shortened official APCO Ten-Signals Code.

10-1 **receiving poorly.**
10-2 **receiving well.**
10-3 **stop transmitting.**
10-4 **Ok, message received.**
10-5 **relay message.**
10-6 **busy, stand by.**
10-7 **out of service, leaving air.**
10-8 **in service, subject to call.**
10-9 **repeat message.**
10-10 **transmission complete, standing by.**
10-11 **talking too rapidly.**
10-12 **visitors present.**
10-13 **advise weather and road condition.**
10-16 **make pickup at ________.**
10-17 **urgent business.**
10-18 **anything for us?**
10-19 **nothing for you, return to base.**
10-20 **my location is ________.**
10-21 **call by telephone.**
10-22 **report in person to ________.**
10-23 **stand by.**
10-24 **completed last assignment.**
10-25 **can you contact ________.**
10-26 **disregard last message.**
10-27 **I am moving to Channel ________.**
10-28 **identify your station.**
10-29 **time is up for contact.**
10-30 **does not conform to FCC rules.**
10-32 **I will give you a radio check.**
10-33 **emergency traffic at this station.**
10-34 **trouble at this station, need help.**
10-35 **confidential information.**
10-36 **correct time is ________.**
10-37 **wrecker needed at ________.**
10-38 **ambulance needed at ________.**
10-39 **your message delivered.**
10-41 **please tune to channel ________.**
10-42 **traffic accident at ________.**

10-43 **traffic tieup at ___________ .**
10-44 **I have a message for you.**
10-45 **all units within range please report.**
10-46 **assist motorist.**
10-50 **break channel.**
10-60 **what is next message number?**
10-62 **unable to copy; use phone.**
10-63 **network directed to ___________ .**
10-64 **network clear.**
10-65 **awaiting your next message/assignment.**
10-67 **all units comply.**
10-69 **message received.**
10-70 **fire at ___________ .**
10-71 **proceed with transmission in sequence.**
10-73 **speed trap at___________ .**
10-74 **negative.**
10-75 **you are causing interference.**
10-77 **negative contact.**
10-81 **reserve hotel room for ___________ .**
10-82 **reserve room for___________ .**
10-84 **my telephone number is ___________ .**
10-85 **my address is ___________ .**
10-89 **radio repairman needed at ___________ .**
10-90 **I have T.V.I.**
10-91 **talk closer to the mike.**
10-92 **your transmission is out of adjustment.**
10-93 **check my frequency on this channel.**
10-94 **please give me a long count.**
10-95 **transmit dead carrier for five seconds.**
10-97 **check test signal.**
10-99 **mission completed, all units secure.**
100-200 **police needed at ___________ .**

OFFICIAL APCO TEN-SIGNALS CODE

This official code adopted by the Associated Public Safety Communications Officers, Inc., is *strongly* recommended for all CB communications where a coded transmission is desired.

10-1 **signal weak.**
10-2 **signal good.**
10-3 **stop transmitting.**
10-4 **affirmative (OK).**
10-5 **relay (to).**
10-6 **busy.**
10-7 **out of service.**
10-8 **in service.**
10-9 **say again.**
10-10 **negative.**
10-11 **___________ on duty.**
10-12 **stand by (stop).**
10-13 **existing conditions.**
10-14 **message/information.**
10-15 **message delivered.**
10-16 **reply to message.**

10-17 **enroute.**
10-18 **urgent.**
10-19 **(in) contact.**
10-20 **location.**
10-21 **call (__________) by phone.**
10-22 **disregard.**
10-23 **arrived at scene.**
10-24 **assignment completed.**
10-25 **report to (meet).**
10-26 **estimated arrival time.**
10-27 **license/permit information.**
10-28 **ownership information.**
10-29 **records check.**
10-30 **danger/caution.**
10-31 **pick up.**
10-32 **__________ units needed specify/number/type.**
10-33 **help me quick.**
10-34 **time.**

BLACK HORSE CODE

This code can be used for privacy or clarity. Numbers can be transmitted as letters or words representing letters. Its origin is uncertain, most likely international short wave, but this code is used extensively in parts of northeastern Canada.

CODE	MEANING
B	**1**
L	**2**
A	**3**
C	**4**
K	**5**
H	**6**
O	**7**
R	**8**
S	**9**
E	**0**

Q SIGNALS

Although the "Q" Code is out of vogue and seldom used today, some of the signals familiar to CBers are listed below.

QRA	what is the name of your station?	the name of my station is __________ .
QRB	how far are you from my station?	the distance is __________ .
QRD	where are you bound and where are you from?	I am bound for __________ from __________
QRE	what is your estimated time of arrival at __________ (place)?	my estimated time of arrival at __________ is __________ hrs.
QRF	are you returning to __________ (place)?	I am returning to __________ .
QRL	are you busy?	I am busy with __________ . please do not interfere.
QRM	are you being interfered with?	I am being interfered with.
QRS	shall I speak more slowly?	speak more slowly.
QRT	shall I stop transmitting.	stop transmitting.

QRU	have you anything for me?	I have nothing for you.
QRV	are you ready?	I am ready.
QRW	shall I inform ________ that you are calling him on Channel ________ ?	please inform ________ that I am calling him on Channel ________ .
QRX	when will you call me again?	I will call you again at ________ .
QRZ	who is calling me?	you are being called by ________ .
QTJ	what is your speed?	my speed is ________ .
QTN	at what time did you depart from ________ .	I departed from ________ at ________ .
QTR	what is the correct time?	the correct time is ________ .
QTU	what are the hours during which your station is open?	my station is open from ________ to ________ .
QTV	shall I stand guard for you on Channel ________ .	yes, stand guard for me on Channel ________ .
QTX	will you keep your station open for further communication with me until further notice or until ________ ?	yes, I will keep my station open for further communication with you until further notice or until ________ .
QSL	can you acknowledge my receipt?	I am acknowledging receipt.
QSN	did you hear me on Channel ________ ?	I did hear you on Channel ________ .
QSO	can you communicate with ________ ?	I can communicate with ________ .
QSX	will you listen to ________ on Channel ________ ?	I will listen to ________ on Channel ________ .
QSY	shall I change to another channel?	change to Channel ________ .
QUA	have you news of ________ (specific CBer)?	here is news of ________ (specific CBer).
QUD	have you received the emergency signal sent by ________ ?	I have received the emergency signal sent by ________ .
QUF	have you received the distress signal sent by ________ ?	I have received the distress signal sent by ________ .
QUM	is the distress traffic ended?	the distress traffic is ended.
QUO	shall I search for ________ ?	please search for ________ .
QUR	have survivors . . . 1. received survival equipment? 2. been picked up? 3. been reached by rescue party?	survivors 1. are in possession of survival equipment. 2. have been picked up. 3. have been reached by rescue party.
QUS	have you sighted survivors or WRECKAGE?	I have sighted ________ at ________
QUT	is position of incident marked?	position of incident is marked by ________ .

TRUCKERS' TWELVE CODE

The Truckers' Code was developed in 1976 by "The Whammers,"a group of gear jammers who were sick and tired of 4-wheelers using their language. Nontrucking CBers are advised to use it at their own risk.

12-1 **one in every crowd.**
12-2 **get your own towel, mine's already wet.**
12-3 **I can't believe I ate the whole thing.**
12-24 **a pill a day keeps the doctor away.**
12-25 **quit chewing the rag.**

12-35 ride on.
12-37 you're jiving me.
12-38 get off my back.
12-39 beats the hell out of me.
12-45 big deal.
12-46 here we are fellas, Miss America.
12-47 beautiful, just beautiful.
12-48 today's just not my day.
12-49 so much for you and the horse you came in on, too.
12-50 if you got it, a truck brought it.
12-61 bull.
12-62 you can't fight city hall.
12-65 screwed up like Hogan's goat.
12-69 excuse please, couldn't find my way.
12-72 you have me confused with someone who cares.
12-73 a terrific screwing up is in progress.
12-74 that figures.
12-75 here I am, Mr. Terrific.
12-76 situation normal and all screwed up.
12-77 want me to call a chaplain.
12-78 oh my God, now he thinks he's a cop.
12-83 who, me?
12-84 can't receive you, a bird messed on my antenna.
12-85 let's be careful and let someone else do it.
12-86 I think you have problems, stupid!
12-94 mobile, you have motor oil in your mouth.
12-96 CB maniacs of America, unite.
12-100 just another cotton picking truck driver.
12-200 don't mess with me.

WARNING: You Are Advised To Use This Twelve Code If And Only If You Are Big Enough Or Fast Enough To Keep Your Own Tail From Being Kicked.

IN THE NAME OF THE LAW

Here are 50 terms used by CBers to describe various law enforcement officials. But if you're on the air and don't want to antagonize a bear with ears (thereby increasing your chances of getting bitten), refer to him with respect—as a highway patrol, police officer, or state trooper.

LOCAL CITY POLICE
blue boy
city kitty
country joe
little bear
local bear
local boy
mickey mitchell
mickey mouse (three-wheel motorcycle traffic cop)
smokey two wheeler

COUNTY POLICE
local yokel
county mountie (sheriff's deputy)
papa bear (sheriff)

STATE TROOPER
barnie
blue jeans
boogie man
honey bear (female)
Kojak
kodiak

pink panther
pole cat
sloppy joe
state bear
whatevers
Yellowstone Parker

LAW OFFICERS EQUIPPED WITH CB

bear with ears
black and white CBer
daddy bear
smokey with ears
sneaky snake

OTHER LAW ENFORCEMENT OFFICERS

baby bear (rookie)
bear barnies (any badge bearer)
bear in the air (police helicopter)
bearded buddy
big brother
blue and white
camera (police with radar)
chicken plucker (highway toll collector)
DOT (Department of Transport; also Canadian Department of Transport)
evel knievel
Friendly Candy Company (FCC)
Fox-Charlie-Charlie (FCC)
green stamp collector
gun runner
jack rabbit
john law
mamma bear (metermaid)
nightcrawler
paperhanger
peter rabbit
picture taking machine
pigs
porky bear
red wheel
salt and pepper
shot
smoke screen
smokey
smokey beaver (any female police officer)
uncle charlie (FCC)

TRUCKER CB ATLAS

Rubber City	*Akron, Ohio*
Eskimo Pie Land	*Alaska*
Big A.	*Amarillo, Texas*
Hog County	*Arkansas*
Hot Lanta—or Hot Town	*Atlanta, Georgia*
Bean City	*Boston, Massachusetts*
Motel City	*Breezewood, Pennsylvania*
Choo Choo Town	*Chattanooga, Tennessee*
Mr. D's	*Chicago, Illinois*
Mile High City	*Denver, Colorado*
Little D.	*Dallas, Texas*
Diesel City or Motown	*Detroit, Michigan*
Bar City	*Forest City, Arkansas*
The Real World	*Disneyland, California*
Cow Town	*Ft. Worth, Texas*
Smoke Stack City	*Gary, Indiana*
Tinsel Town	*Hollywood, California*
H-Town	*Hopkinsville, Kentucky*
Hot Water City	*Hotsprings, Arkansas*
Astrodome City	*Houston, Texas*
Circle City	*Indianapolis, Indiana*
Bugger Hole City	*Irving, California*
Capital J.	*Jackson, Mississippi*
J. Town	*Jackson, Tennessee*
Finger Lickin' Country	*Kentucky*
K. Town	*Knoxville, Tennessee*

K.C. Town	*Kansas City*
Dice City (or) Divorce City	*Reno, Nevada*
Rock City	*Little Rock, Arkansas*
Shakey City	*Los Angeles, California*
Derby Town	*Louisville, Kentucky*
Piggybank State	*Maryland*
Big M. or River City	*Memphis, Tennessee*
Tell-Me State	*Missouri*
Magnolia State	*Mississippi*
Beer City	*Milwaukee, Wisconsin*
Monkey Town	*Montgomery, Alabama*
Nastyville or Guitar City	*Nashville, Tennessee*
Default City or Big Apple	*New York City*
Superdome City	*New Orleans*
The Rock's Park	*New York State*
Dirty Side Center	*Newark, New Jersey*
Buckeye State	*Ohio*
Rose City	*Portland Oregon*
God's Country	*Oregon*
Power City	*Pittsburgh, Pennsylvania*
Cactus Patch	*Roswell, New Mexico*
Dick's Trick	*San Clemente, California*
Friskie City	*San Francisco, California*
Sport City	*Shreveport, Louisiana*
Cigar City	*Tampa, Florida*
Big T.	*Tucson, Arizona*
Watergate City	*Washington, D.C.*
Hole in the Wall	*Wheeling, West Virginia*

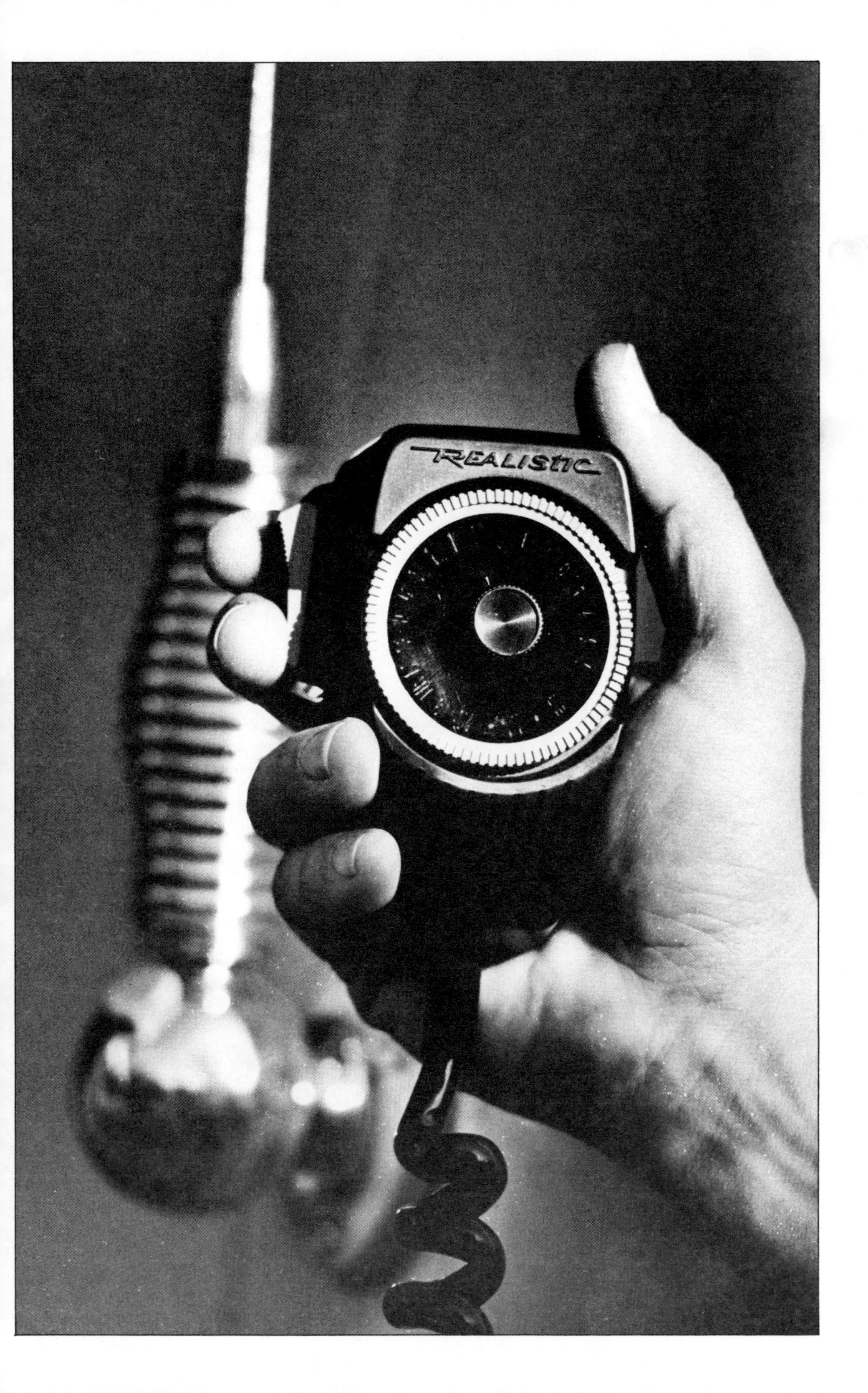
REALISTIC

4.

Rebuilding the Tower of Babel: CB and the Federal Communications Commission

Uncle Charlie is Swamped by 20,000 License Applications a Day

Forty sacks of mail sat ominously inside the drafty front door of the converted new-car dealership in Gettysburg, Pennsylvania. Outside, the March winds howled unmercifully. Promptly at 9:00 a.m., fifty-one beleaguered employees of the FCC Special Facility (which handles CB and other radio license applications) trooped through the door, punched in, doffed their heavy winter coats and took their places along an assembly line of desks. Another week had begun and another 100,000 CB licenses would be processed. The facility had been swamped for months (room had to be found even in the ladies' lounge to store stacks of correspondence) and the flood showed no sign of abating. FCC officials expect their Gettysburg facility to handle nearly 4 million new CB applications during 1976—almost as many as have been processed during the entire 18-year history of the Citizens Radio Service.

Early in 1975, following the Arab Oil Embargo, the subsequent 55 mph national speed limit, and the nation-wide, CB-orchestrated truckers' shutdown, applications began to rain down upon the Gettysburg facility. In January, 1975, for example, 72,688 applications were received, 124% more than the January, 1974 total. The FCC staff appeared to be bending under the weight of the work this sudden, surging interest in CB radio. The Commission doubled its work force in a futile effort to keep abreast of the flood. Ninety percent of the applications are received clean — that is, ready for issuing, but an outside contractor had to be brought in last year to keystroke the information from the application onto a disk to speed up processing in the commission's computer.

Nevertheless, it still took the FCC up to ten weeks to process a clean application throughout most of 1976. And that delay added to the commission's woes as eager would-be CB operators deluged the FCC with telephone calls and letters asking the status of their applications. (So busy are the Commission's telephones that Pennsylvania Bell reports only 20% of the calls were even getting through—those that did numbered about 400 per day (including about 150 per week from Congressmen with constituent complaints). As for the mail, the Commission faced the prospect of answering more than 20,000 pieces daily. In an effort to reduce this mounting backlog of CB license applications, the FCC introduced, in September, 1976, a self-mailer licensing system which it hoped would eventually reduce processing time to little more than one week. Under the new system, manufacturers include in all CB-radio packages a two-part license form. Half of the form is to be filled out by the applicant and mailed to the FCC, for processing by the Commission's new high-speed computer. The other half is to be retained as a temporary license, good for 60 days, enough time these days to have a permanent license processed. Any operator who chooses not to send in his license application runs the risk of incurring a $10,000 fine or up to a year in jail or both.

Even with the improved processing system, some 25–30% of all CB operators never get around to applying for their license.

The FCC's 100-man CB enforcement squad and small budget is not adequate to properly police this situation. Adding to the Commission's problems are the results of a survey made in 1975 by the government's General Accounting Office. A sample of 3,740 business users, whose CB licenses were due to expire, indicated that 31% did not intend to renew, in spite of the fact that most of them required CB-radio service for business purposes. The Commission said 80% of these legitimate users named idle chatter—supposedly banned for periods exceeding 5 minutes at a stretch—as their principal reason for not renewing the licenses. (In 1975, the FCC received 31,000 complaints about CB violations, including 250 direct complaints from members of Congress. Most were for interference and illegal broadcasting.) CB violations, in fact, now constitute more than half of all complaints for any radio or television service governed by the FCC.

FCC CLASS-D LICENSE APPLICATIONS 1958–1976

Fiscal Year (July 1)	Licensed Stations	Application Receipts
1958	38,611	5,276
1959	49,269	25,346
1960	126,034	106,530
1961	206,106	140,153
1962	305,135	171,984
1963	446,590	221,332
1964	682,307	293,480
1965	744,613	210,325
1966	795,502	223,845
1967	848,237	223,761
1968	867,552	216,629
1969	860,624	234,957
1970	886,951	229,435
1971	868,013	163,569
1972	851,610	185,454
1973	836,924	230,804
1974	931,545	341,452
1975	1,245,681	574,904

Most violations involve unlicensed operators, CB operators failing to properly identify themselves (via FCC-assigned call letters), and interference with other radio and television users, caused by illegal, overpowered CB transceivers.

The GAO report said that despite the fact that CB communications are restricted to low-powered transmitters on assigned channels, CB users have been detected "using 84 different frequencies assigned to government agencies and other mobile-radio users and using high-powered transmitters which frequently interfere with television reception and other electronic facilities."

Requesting information from state officials on their own experiences with questionable or illegal use of the CB, the GAO also found that 42 states reported the use of CB to violate laws or to avoid law enforcement. Forty-one states reported the use of CB to avoid speed, weight, and/or road licensing checks. Seven states, principally California, Illinois, and New York, reported the use of CB for burglary, robbery, drag racing, unlawful assembly, prostitution, narcotics trafficking, smuggling, and transporting stolen goods. Not to mention the two million or more unlicensed operators who flagrantly violate the FCC every time they go on the air. Twenty states considered CB misuse a serious law enforcement problem and twelve have asked for more FCC enforcement. Several bills pending now before Congress may insure this.

C. Phyll Horne, the FCC's top enforcement officer, faces a special problem. Hampered by budgetary, legislative, and jurisdictional limitations, he finds it almost impossible to police CB. "It's not even a federal crime to attack an FCC official," says Horne, whose field agents have been set upon by irate CB rule-breakers with increasing regularity.

Is this what the golden age of communications has come to? A lawless mob attacking federal inspectors, mouthing obscenities over the air, and using CB to plan crimes and evade the law? "Hardly," says Horne, who can call on staff in thirty offices around the country to enforce the rules. In the past year, Horne's Field Operations Bureau has added four eight-man mobile teams which are stationed at Laurel, Maryland; Atlanta, Georgia; Grand Island, Nebraska; and Santa Ana, California, but who range the entire country attempting to track down violators.

FCC'S SIX MOST SERIOUS VIOLATIONS

1. **Failure to use call signs at beginning and close of transmission.**
2. **Use of a linear amplifier (or any illegal power booster).**
3. **Use of obscenity, foul language, or advertising.**
4. **Intentional use of CB to violate state or local laws.**
5. **Skip talking (intentional transmission beyond the legal 150 mile limit).**
6. **Any transmission outside the 27 MHz band assigned to the Citizens Radio Service.**

The Candyman Strikes Back

"We have some new and very sophisticated equipment," Horne says, "and we can catch a mobile operator in violation of FCC rules as quick as we can zero in on a permanent base station." First offenders, Horne points out, usually receive a $50–500 fine and loss of license. But the second offense is a federal felony, with a $10,000 fine and up to a year in jail. "We don't have to use a triangulation (direction finding) system anymore," Horne adds. "We can home right in on our target, and we'll use trucks, cars, and helicopters if we have to."

In fiscal year 1975, the FCC issued 1,363 notices of apparent liability, lifted more than 200 licenses for persistent offenses, and issued over 200 cease and desist orders to licensed and

unlicensed operators (about four unlicensed operators per week, on average, actually go to jail for illegal operation of a CB station). Most offenders straighten up, but there are maverick operators who persist in the belief that they own the airwaves. As one safety and special staffer noted wryly, "CB is certainly a weird environment."

Besides these efforts at solving or at least easing some of the problems it now faces, the Commission conceded earlier in 1975, that the Citizens Radio Service could no longer be contained within the channels then assigned to it. Accordingly, CB was expanded from the original 23 channels to nearly double that number in mid-1976, with the possibility that another 20 channels might be made available in the near future.

All of which would seem to put to rest to doubts anyone might have about the extent to which citizens radio might be used. Charles Higginbotham, Chief of the Safety and Special Services, was with the FCC back in 1958 when the Citizens Radio Service was inaugurated. He's also an ardent CBer himself. Higginbotham likes to say that "The uses of citizens radio are as broad as the imagination can conceive." He expects "performance in reasonable compliance with the rules," but exemplifies the FCC's tolerant attitude toward this most democratic of communications systems. Many once feared that the early dream of universal citizens radio would become a Tower of Babel. Under the sympathetic leadership of men like Charles Higginbotham and Phyll Horne, the FCC has shown CBers how accessible and effective owning and operating your own radio station can be.

5.

The Basic Hardware: CB Equipment and Accessories

How Radio Works

Radio is based on the movement of *energy* through *space*, but since we can't see either, it's easier to understand the basic principles if we think in terms of materials like water. If you were sitting in a boat in the middle of a lake and you slapped the surface of the water with the palm of your hand, you would generate ripples or waves that would travel outward from the point of impact, moving in concentric circles of increasing diameter, until they reached the shore. The number of waves generated each second is the *frequency*. The size of the waves, measured from the crest of one to the crest of the next, is the *wavelength*.

Notice that frequency and wavelength are inalterably related. If you think of the work done to generate the waves (your hand slapping the water), you should realize that in order to make bigger waves, you have to push deeper into the water, and that takes more time than is needed to make smaller waves. Long wavelengths correspond to low frequency (remember long goes with low) and short wavelengths correspond with high frequency.

Radio waves can move through water, or many other substances, but usually travel through air or a vacuum. Instead of a rolling wall of water, radio waves consist of a moving field of electromagnetic energy (electric and magnetic forces) that can't be felt, but can be measured, or converted to audible energy by a radio receiver.

Radio transmitters are much simpler than receivers. A basic unit consists of a vibrating *crystal* in an "oscillator" stage that generates an electrical signal of specific frequency. This unit is then *modulated* by the signal representing your voice that comes through a microphone. After modulation, the electronic signal is raised in frequency ("multiplied") and finally amplified and fed to the antenna, which forms waves in the air in much the same way as your hand forms waves in water. The size of the antenna is carefully calculated with the size of the wave to be generated.

A radio receiver uses a similar antenna to pick up transmitted waves and its circuitry works in the reverse order of the transmitter. The antenna feeds the received waves into an amplifier. The signal is then demodulated extracting the signal from the carrier wave and lowering its frequency before it is amplified and fed to a speaker. The speaker works in the reverse fashion of a microphone, and changes electrical energy to physical energy—the sound. Receivers are generally much more complex than transmitters (and more expensive to build) because there can be many stages, each designed to home in on the specific desired signal, and to reject all other frequencies. The transmitter just generates a signal and pushes it out.

Radio communication is possible over a very wide range of frequencies (and corresponding wavelengths) with the specific frequency being selected for desired range, freedom from external interference, and other technical reasons. CB works at about 27 MHz, which is higher than regular AM broadcasts and lower

If you slap the surface of water, the pressure from your hand will generate waves. Wavelength is the distance from the crest of one wave to the crest of the next. Frequency is the number of waves generated in a particular time.

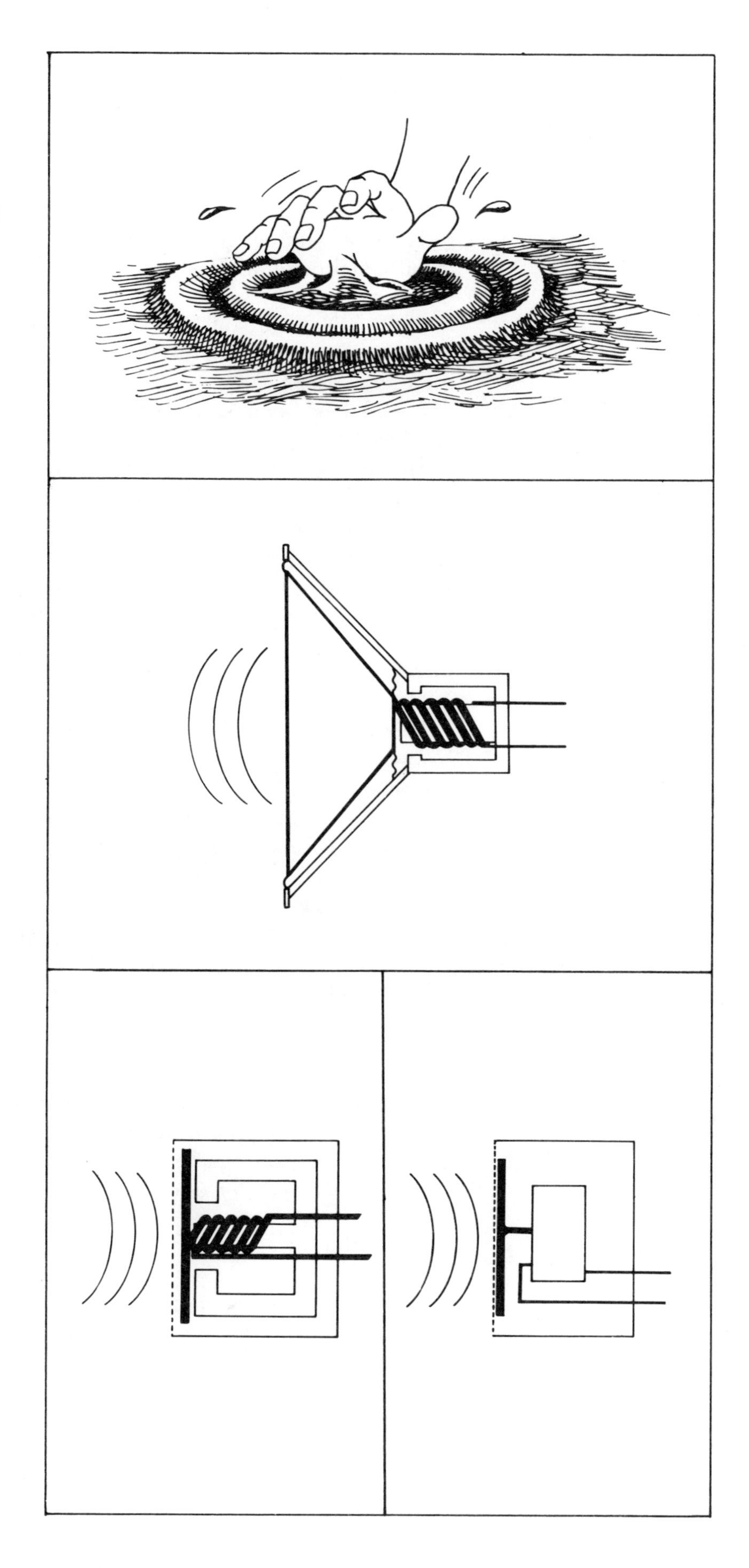

A speaker has a paper cone or "diaphragm" attached to a coil of wire suspended near a permanent magnet. When electricity is fed into the coil of wire, it moves the cone away from the magnet and sound comes out.

A dynamic microphone works in the reverse fashion of a speaker. When the diaphragm is struck by sound waves, it moves a coil near a permanent magnet and puts out electrical signals.

A crystal or ceramic microphone generates electric impulses when sound waves press a diaphragm against a piezoelectric material.

than FM or television.

CB Hardware and Accessories

CB transceiver performance varies only a bit. Transmitter output power is limited by law, and although extra dollars can buy a better receiver section, the big difference among CBs is in appearance, convenience, and features.

There are three general classes of CB transceivers: base, mobile, and portable. Base units, designed to be used at home or in a fixed business location, are connected to a regular 110 volt–AC wall outlet, and tend to be larger than mobile CB units. Some of the extra space holds useful features, but a lot of it goes for frills and cosmetics. It is not unusual for a base unit to sell for $60 – 80 more than a comparable mobile, with the only real difference being the *power supply*.

The mobile unit is, by far, the most common CB transceiver, usually compact and designed for connection to the 12-volt DC power available in cars, boats, and small planes, and usually operable on all legal channels. Any mobile can be used as a base unit with an accessory power supply, available for $10–30, and many accept battery packs for portable use.

Carry It With You

True portables, or "walkie talkies," vary considerably. There are small, inexpensive units, available from Radio Shack, Lafayette, and others, that sell for as little as $6.00, put out less than one milliwatt (1/10 watt), and do not require licenses. Most of these are children's toys, seldom reach more than a few blocks, and operate on only one or two channels. One serious new unit, the Pocket-Com, is tiny (smaller than a pack of cigarettes) and yet electrically very sophisticated. Its limited range makes it best suited to local "paging." There are full-power multi and all-channel 4-watt units made by Midland, Lafayette, Radio Shack, Zodiac, Handic, Pace, and others that weigh several pounds and offer most of the features found in regular base or mobile units. You can also choose from many intermediate portables, with from 1 to 3 watts, each offering various features at various prices. Most walkie talkies can be connected to an AC or 12-volt adapter and have external antenna connections for use indoors or in cars.

Sidewinders

CB base or mobile units are available to work on either AM or Single Sideband (SSB). SSB is a more efficient means of communication that uses less frequency space ("bandwidth") to carry the signal, and less power, usually reaching twice as far as AM-CB at the legal power limit. There is an upper and lower sideband for each channel, and it is possible for one conversation to be going on on the upper sideband and another on the lower with no interference.

A conventional AM-CB transceiver uses both sidebands, and if an AM station is transmitting, no sideband operation is possible on that channel. You will often see SSB sets advertised as having 69 or more channels, but they can only have, for example, either 23 (AM) or 46 (SSB) channels, because AM and SSB units cannot operate in the same space at the same time. SSB sets seem to be the wave of the future. Single Sideband makes better use of the frequency spectrum, talks farther, and because SSB receivers pick up a narrower range of frequencies, they pick up less noise than AM sets. But the benefits will come costly. Because of the more complex circuitry involved, there is usually a $100 or so difference between a manufacturer's top AM set and his bottom SSB device. SSB transceivers are often larger than AM-only units, and a bit more complicated to operate. And if the channels in your area are already clogged with AMers, SSB won't really offer much of an improvement.

The people on SSB (called sidewinders) are a breed apart from regular CBers. They consider the use of Ten Codes part of a juvenile "Highway Patrol"/"Adam-12" syndrome that actually slows down communication rather than giving it efficiency. Sidewinders are more likely to have handles like Howie or Pete than Red Dog or Hatchetman, and use real English, like "hello," instead of CBese, like "Breaker by break, break, break." Sidewinders are usually polite and knowledgeable about their expensive equipment, very serious about radio, and they obey the FCC regulations. Regular AM CBers would do well to follow their example.

What Those Knobs and Switches Do

CB transceivers range from $50 to $700 or more, and the price indicates features and flexibility rather than major differences in performance. If you only need CB for emergency use, a Channel 9-only unit can be had for as little as $39 from some mail order houses. A six-channel unit like the Realistic TRC-11 ($80) offers a bit more talking room, and makes more sense. If you only want to *listen* to Smokey reports or suffer from incurable mike fright, converters are available to allow CB reception on your car AM radio. Tenna makes a good unit; the model CBC-23 lists for $35.

But the chances are you're going to want to enjoy the complete CB experience. And that means two-way communications. So, to help you on the air, let's take a minute or two to learn what those mysterious knobs and switches on your new CB transceiver are for. The first three are found on every CB, the others on some.

VOLUME CONTROL. As on any other radio, it varies the loudness of what you hear and usually turns the set on and off. Remember to lower the volume before plugging in an earphone.

CHANNEL SELECTOR. Chooses the channel you will transmit and receive on. Most 23 channel CB sets use a rotary knob with lighted digits. Rigs equipped for both the old and the newly-opened channels may use two knobs (like Palomar's Digicom 100) or even a calculator-style keyboard (SBE). In a base unit this switching device doesn't make much difference. In a mobile, simpler ones are safer, since you won't have your eyes off the road for so long. Radio Shack's "One-Hander" and the Royce 1-580 have the channel selector conveniently located right in the microphone.

SQUELCH. This knob lets you keep the transceiver silent unless a signal of a preset level is present. It eliminates background noise, and cuts out weak stations if you want.

RF GAIN CONTROL (or local/distance switch). This can be used to "desensitize" your receiver if a transmitter is so close to you that it is coming in distorted.

MICROPHONE GAIN CONTROL. Controls a circuit that boosts the output of your microphone to provide maximum modulation for strong output. The better circuits also limit the maximum modulation to avoid distortion, and may keep the volume constant as you move the mike closer to and away from your lips.

DELTA TUNE. A knob or switch that lets you tune slightly off frequency if the transmission you are receiving is itself slightly off that frequency. You'll hardly ever use it for this, its intended purpose, but it sometimes comes in handy to get away from someone bleeding over from another channel.

CLARIFIER. A control found only on SSB units, used for fine tuning. Difficult to adjust while driving.

MODULATION INDICATOR. A light that flashes in time with your words to indicate relative output strength and to show that you are getting out, (actually transmitting).

S METER. Measures the strength of incoming signals. Unnecessary, but almost standard equipment. Lets you tell others how well they are putting out. S-meter calibration varies greatly and is somewhat arbitrary with each manufacturer. Two CBs in the same location and connected to the same antenna could give widely varying measurements of the same incoming signal. Meters vary in size from the thumbnail-size unit found on walkie talkies to the "professional" size used on the Panasonic RJ3200 or Craig 4103.

Big ones take up valuable space in mobile units, and to some degree are a styling gimmick, but they can be more accurate, and easier and safer to read.

RF METER. Usually combined with an S meter. Functions during transmission and gives you an idea of your output strength. It is often calibrated in watts, but the calibrations are usually inaccurate.

SWR METER. Until recently only available as an accessory add-on unit, this useful device is now appearing in larger base units, such as the Realistic Navaho TRC57 and Teaberry Stalker II. It indicates how well your antenna is matched to the transmitter and lets you know if power is being wasted.

TONE CONTROL. Available in models from Panasonic and Regency, among others, this control sometimes helps to eliminate interference and may provide better sound when an external speaker is used.

EXTERNAL SPEAKER SWITCH. Sends the received signal to a speaker mounted away from the CB, perhaps in another room, or outside your car. Can also be used to feed an accessory speaker near the transceiver, if the one in the set sounds tinny.

PA SWITCH. Lets you use the CB as a public address amplifier when connected to a remote speaker. Works with your regular microphone.

ANL. Automatic Noise Limiter, cuts out noise caused by automobile engine and other sources. To some degree it lessens receiver sensitivity, but this will seldom be a real problem.

NB. Noise blanker, another type of noise suppressor, sometimes better than ANL, sometimes not noticeable at all. It's effectiveness depends on your car's electrical system.

CLOCK. Handy on a base station, if there's no other clock nearby. Sometimes connected to a timer, as on the Lafayette Telsat 925, that can be set to turn your CB on at a predetermined time. May be either an inexpensive "flipover" digital unit or a super accurate electronic digital device, as on the Radio Shack Navaho TRC57.

PILOT LIGHT. Shows that your set is turned on, and power is connected. May be a separate indicator or the tuning dial illuminator.

TRANSMIT LIGHT. Turns on when you push the mike button. Can be a simple red light, as on the Cobra 29, an elaborate "On the Air" read-out like the Gemtronics GTX3000, or the S-meter illumination may switch from white to red, as on the Midland 13-867 or Radio Shack Navaho TRC57.

MICROPHONE JACK. Most CB microphones plug into a jack. Side jacks, such as used on the Cobra 138, are less common than front jacks, but waste less space and are safer mobiling than are front jacks. CBs with permanently connected mikes, rather than plug-in jobs, are harder to trouble-shoot and more costly to repair, but are less likely to get disconnected. The best approach is a mike connector that screws as well as plugs in.

CHANNEL 9 MONITOR. A second receiver within your transceiver, permanently tuned to the emergency frequency. Usually flashes a light when there is a signal present on 9, or may automatically turn on 9, killing other channels. Useful for REACT bases and other emergency teams.

WEATHER-CHANNEL MONITOR. Found on the Pace CB145, Lafayette HB700, and others, this is a separate receiver circuit that is permanently tuned to receive federal government weather reports, available in most coastal areas. Particularly important for boaters, and a boon for any traveler.

HANDSET. More and more CB-set makers (Teaberry, Johnson, Radio Shack, Lafayette, and Midland, for example) are offering gear with telephone type handsets. It's a gimmick, but has serious advantages: it forces you to hold the mike close for good output; it keeps the incoming signals from disturbing others in the house or car, and it gets the sound close to your ear so it won't have to compete with wind, engine, and other sounds. Transceivers with handsets also have regular speakers that can be used when you are monitoring and don't want to hold the handset to your ear or when other people want to listen. If your CB did not come with a handset, Hy-Gain and Lafayette make add-on units that will work with most CBs.

VOX. Voice Operated switching. Automatically switches from receive to transmit when you speak into the mike. A great safety feature for mobile units, but usually found only on large base units like the Tram D201.

Anatomy of A Transceiver

There are, of course, some differences behind the front panels of the CB transceivers, but with few exceptions they are minor. Some "manufacturers" are distributors who buy from other manufacturers and have their own name plates or front-panel design put on stock items. Many companies design equipment that others produce. A few maverick firms actually turn out some of the most interesting and technically advanced equipment, as well as a few "dinosaurs" that somehow are competitive.

The latest circuitry advance relates to how the transceiver frequencies are set. In the old days, CBs had one crystal for receiving and one for transmitting on each frequency. That was expensive and space-wasting enough with 23 channels, but becomes an impossibility when the number of channels is increased. A number of manufacturers, have adapted the phase-locked loop circuit used in aerospace technology and high fidelity, and digital computer devices to product CBs that require only one or two crystals to work on 23, 50, or possibly several hundred frequencies (should FCC regulations change again). Only a small integrated circuit has to be switched, and the set is reprogrammed.

On the other hand, we have sets from Browning, Sonar,

Gemtronics, and Tram that still use vacuum tubes instead of transistors or integrated circuits. You might think they're obsolete—you have to wait for them to warm up, occasionally a tube burns out, only a few of them will fit in today's small cars—but the people who own them love them, and they transmit every bit as clearly as the newest solid-state rigs.

CB specifications are fairly easy to understand and evaluate. Here are the most important ones:

RECEIVER SENSITIVITY. Refers to the ability to pick up stations. Smaller numbers are better. A typical rating is 0.5 μV (microvolts).

ADJACENT CHANNEL SELECTIVITY. This refers to the receiver's ability to ignore signals present on "next door" channels. The higher the number, the better, 50 dB is typical.

TRANSMITTER OUTPUT. Relates to how far your signal will carry. FCC limit is 4 watts on AM and 12 watts on SSB. Some manufacturers boast of 5 watts, this is *input* to the final stage of the transmitter, not actual output.

POWER REQUIREMENTS. Base units are set up to work on about 120 volts AC, mobiles use 12 volts (sometimes listed as 13 or 13.68) DC.

AUDIO OUTPUT. Refers to power available to drive the speaker. One watt is enough for the CB's own speaker; for PA use, you will want 3 to 6 watts.

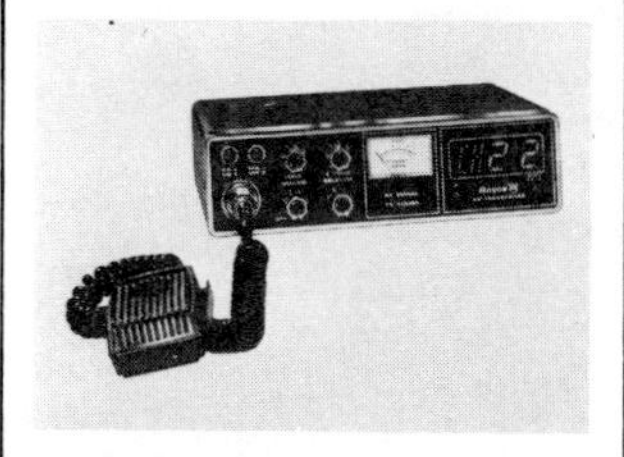

ROYCE 1-610. A boon for any nightrider who has ever squinted across a dark, smoke-filled car, this CB transceiver uses huge electronic digital readouts to indicate the channel. A button on the mike lets you skip from frequency to frequency. Other features include a protective circuit (that flashes a warning light and prevents transmission of SWR is too high or the antenna cable is broken), tone control, delta tune, ANL, and p.a. function. Suggested retail price is $270.

JOHNSON MESSENGER 132. One of the most handsome base units made anywhere, the 132 is housed in the same shell used for Call Director telephones, and it blends into any office or living room where more "technical-looking" sets would be banished. It's a good performer, has a handset for clarity, privacy and noise control, and also functions as a PA amplifier in business or industrial applications. You can have the sound coming from the handset alone, or both the handset and the transceiver. Suggested retail price $260.

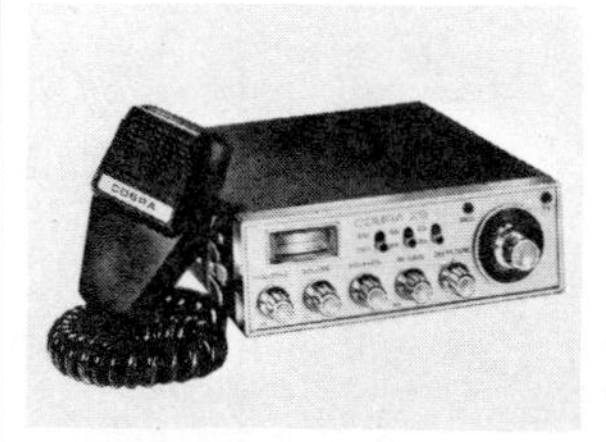

COBRA 29. This transceiver has a long-standing reputation as a powerful performer. In addition to the usual transceiver features, it includes a microphone preamplifier and RF gain control to match your transmitting and receiving power to the station you are working. There are also lights to indicate transmit mode and to show relative modulation, and a nice PA circuit controlled by the mike-level knob. Suggested retail price is $220.

The Lafayette HB-700 is a combination CB transceiver and VHF-FM receiver. It has two weather station frequencies and can also accept a crystal tuned to the police, fire department, marine telephone, or other radio service. It is particularly rugged, with a splash-proof cover and corrosion-resistant construction for boat installations. The speaker is mounted on the top, unlike most mobile CBs, so it works well when sitting on top of a shelf or cabinet. Price is $200.

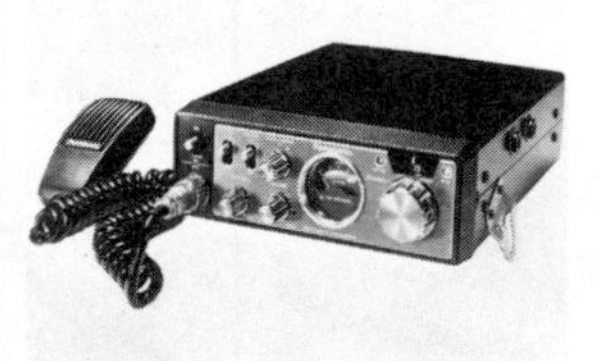

The Panasonic RJ-3200 is simply a terrific transceiver. On the receiving end, it sounds better than any mobile unit we tested, and it has the most consistently high output across all channels of any mobile unit we've seen. The set is very attractive, and extemely well designed. The S/RF meter is large and multicolored for quick, accurate readings. The ANL and noise blanker use one switch—a space saver, and really the most logical approach. A tone control works as a pretty effective noise filter. Pilot lights indicate "on the air" status and show relative modulation strength. Delta tuning is much more effective and useful than on most units we've tested. It's loaded with extra touches—like a heavy-duty power connector, extra-versatile mounting bracket, and a forward tilted speaker "belly" that keeps the sound out of your socks—that clearly shows that a lot of time was spent in its design. Suggested retail price is $180.

A good way to sample CB without making a major investment is with a CB converter, such as the Tenna CBC 23, shown here. The converter connects to your AM antenna, and a short cable goes from the converter to the AM radio. You use a switch to select either AM or CB, and tune in the AM radio to pick up the CB channels, following the channel chart on the front of the converter. If you like, you can set one or more of the automatic pushbuttons on front of the AM radio to instantly tune in a particular CB channel, such as 19 or 9. We found the CBC 23 not as selective as a "real" CB, but adequate for most highway and suburban use, and the sound quality was quite good. Suggested retail price is $35.

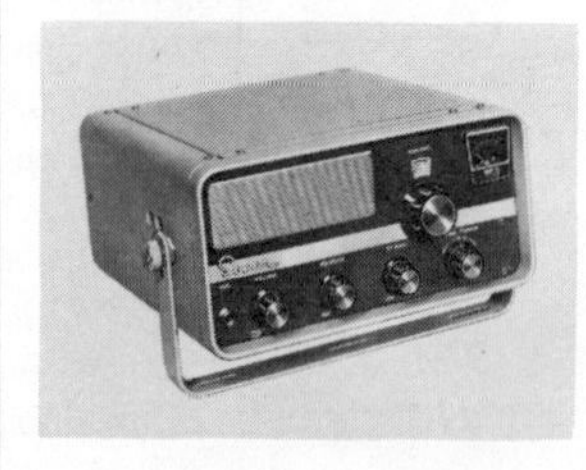

The Sonar FS-3023 is a hybrid design, using both tubes and transistors. It's bigger than all-solid-state CBs, and draws more electric current, but performance is tops. Output power was the highest of any unit we tested, and was quite consistent across all channels, no matter how the antenna was tuned. The receiver section is particularly sensitive for long-range communications, and good selectivity combined with a well-designed fine tuning control makes it a good bet for close-range contacts in places where the channels are packed. It's very rugged, built like a marine or ham transceiver, and conservatively designed to keep operating for a long, long time. Suggested retail price is $395.

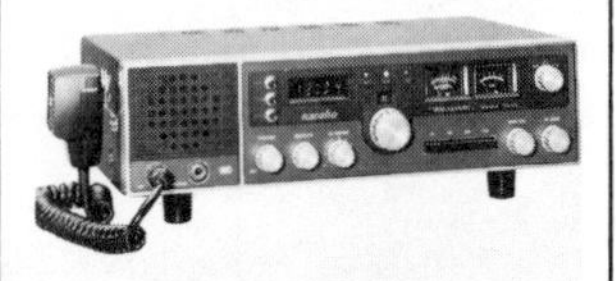

The Realistic Navaho TRC-57 is one of the finest rigs we've seen. Performance is great, and it's a real pleasure to use—it's loaded with features, and every control is sensibly placed and easy to handle. It gives you the feeling that you're a professional radio operator. Adjacent channel rejection is unusually high, audio quality is tops, both coming in and going out, and the unit was completely free from TVI. Features include an accurate electronic digital clock that can be set to operate on the conventional 12-hr time base or on the 24-hour system, pilot lights that indicate AM, or upper or lower sidebands, a clarifier control that also works very well for fine-tuning AM stations, a built-in SWR meter for

keeping tabs on your cable and antenna, front panel headphone jack, PA, ANL, automatic mike gain control, and noise blanker. Suggested retail price is $400.

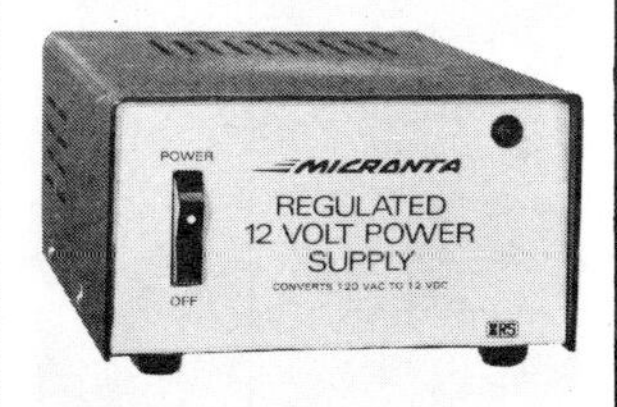

A power supply such as this unit from Radio Shack converts your home's 120 volt AC electricity into 12 volts DC, so you can use your mobile rig indoors. It sells for $26, saving you a good chunk of the $50–80 difference between most mobile and base CB transceivers.

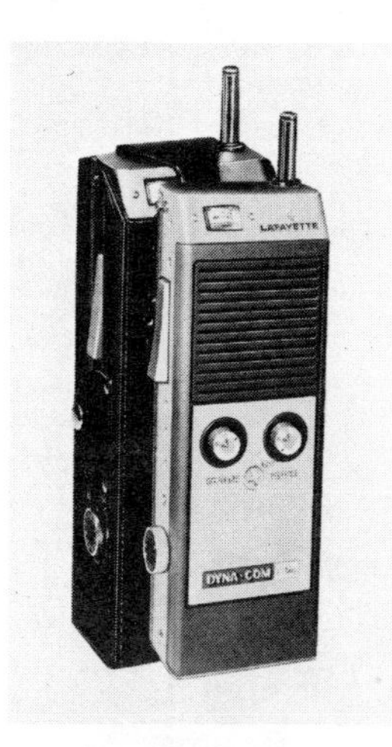

Lafayette has a reputation of building some of the best walkie-talkies. Their rugged Dyna-Com models have survived the Arctic, the desert, the jungle, and are used by many auxiliary and small-town police forces as well as private guards and CB emergency teams. One of the authors of this book has a 6-yr-old Dyna-Com, that's been used for hundreds of hours and has been dropped a dozen times, with no attention other than charging the batteries. The Dyna-Com 12A shown here has a full 4-watts output, and works on 12 channels. It has a meter that indicates battery strength as well as the power of incoming and outgoing signals, connections for a hand-held microphone and external antenna and power source, a squelch control and other features. Price is $110.

The Regency CR-240 is particularly good electronically, having about the best selectivity we've seen, plus a very effective noise blanker circuit. It has a wealth of operating features that will be very useful for the boatman or truck driver. These include a high-power "hailer" PA system, a fog horn, and an alert horn that can be fed to a PA speaker. The meter is large and easy to read, and there are switches to select ANL and distant or local reception, plus a high cut filter to reduce static. Suggested retail price is $240.

The Kraco KCB-2390 is a CB transceiver, an AM radio, an FM stereo radio, and an 8-track stereo tape player . . . all in one housing not much bigger than a standard AM radio, and designed to fit in the hole in the dashboard where the AM set goes.

This means that you don't need three or more different boxes under your dash, you eliminate a lot of wires, it's hard for anyone to tell that you have a CB, and harder to steal if someone does notice it.

The unit is an excellent performer, although selectivity is just a couple of hairs lower than many conventional CBs. It's great to be able to have the CB transmissions coming over the stereo speakers, and the broad-ranging tone control makes it easy to filter out static and adjust for the best vocal sound.

It has all the usual radio and tape features, including tape program indicator lights, stereo indicator light, and local/distant switch, plus the important CB controls, and one very unusual feature. A standby switch lets you listen to tape or regular radio, and it automatically switches to CB if a transmission comes through on the channel you selected.

It's easy to install; and the control shafts can be shifted to accommodate a wide range of dashboard designs. Suggested retail price is $360.

The Midland 13-884 is a mobile transceiver equipped for handset operation. It can be mounted on the transmission hump, or under the dash like a conventional CB. A built-in SWR meter and matching network let you tune the antenna without

extra test equipment or "matchboxes." Other features include an RF gain control to cut back the sensitivity when you are working a nearby station, jacks for external and PA speakers, transmit mode indicator light, ANL, hi/lo tone switch, and positive or negative ground operation. The incoming signal can be fed to the handset only, the handset and the transceiver speaker, or to an external speaker. Suggested retail price is $205.

PACE CB145. A fine, all-around mobile rig, the CB145 is particularly valuable for boaters and long-distance travelers because it includes a separate receiver to pick up weather broadcasts provided by the federal government from transmitting sites around the country. These forecasts are highly detailed and revised frequently, and can make the difference between a pleasure drive (or sail) and a washout. Other features of the CB 145 include transmit and receive mode indicator lights, noise blanker, and p.a. *Suggested retail price is $210.*

PALOMAR DIGICOM 100. One of the most technically sophisticated units available, the Digicom leaves off the frills to concentrate on performance. The unit uses advanced frequency synthesis to generate all channels without using a pile of crystals, and can be easily modified to work on 50 or even 100 channels by switching a small integrated circuit whenever the FCC changes the rules. It uses plug-in circuit boards for easy servicing, and has a noise blanker and public address function. It works beautifully on either AM or SSB, and while designed for mobile operation, it is really too deep to fit under the dash in most small cars. Suggested retail price is $500.

REALISTIC CB-FONE 23. This attractive mobile unit features telephone-type handset operation that gets the sound to your ear without picking up much outside noise, keeps the CB sound from bothering others, and positions the microphone for strongest modulation. It can be switched to work with the speaker in the transceiver as well as in the handset, and an external speaker can also be connected for CB or PA use. The noise blanker is more effective than on most sets, and the delta tune the best we've ever used. Suggested retail price is $180.

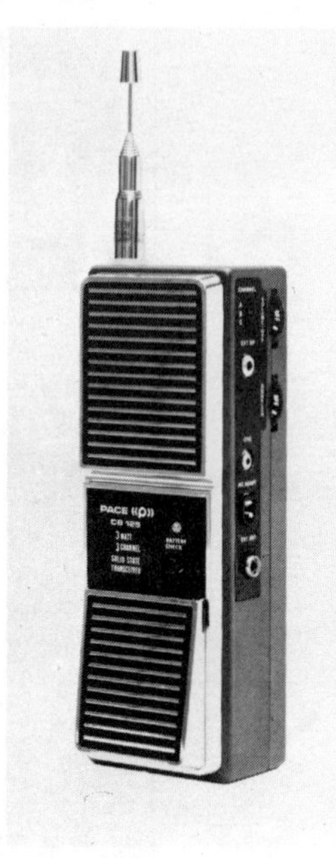

PACE CB 125. An intermediate-type walkie talkie, this unit puts out 2.5 watts, and requires a regular CB license. It works on three different channels, has a squelch control, and connections for an external speaker, external antenna, and AC adapter so it can be used as a base-station transceiver. Suggested retail price is $65.

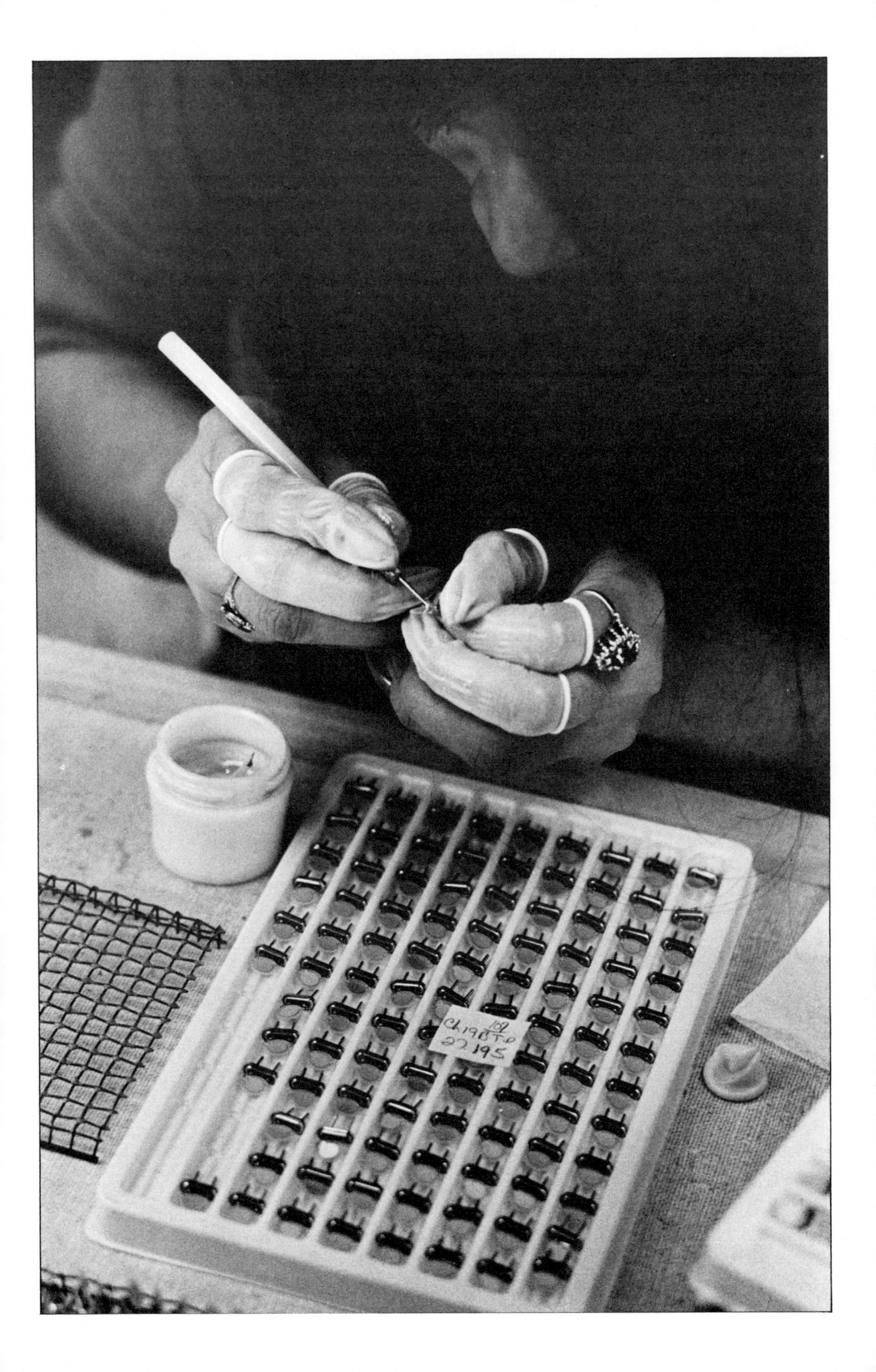

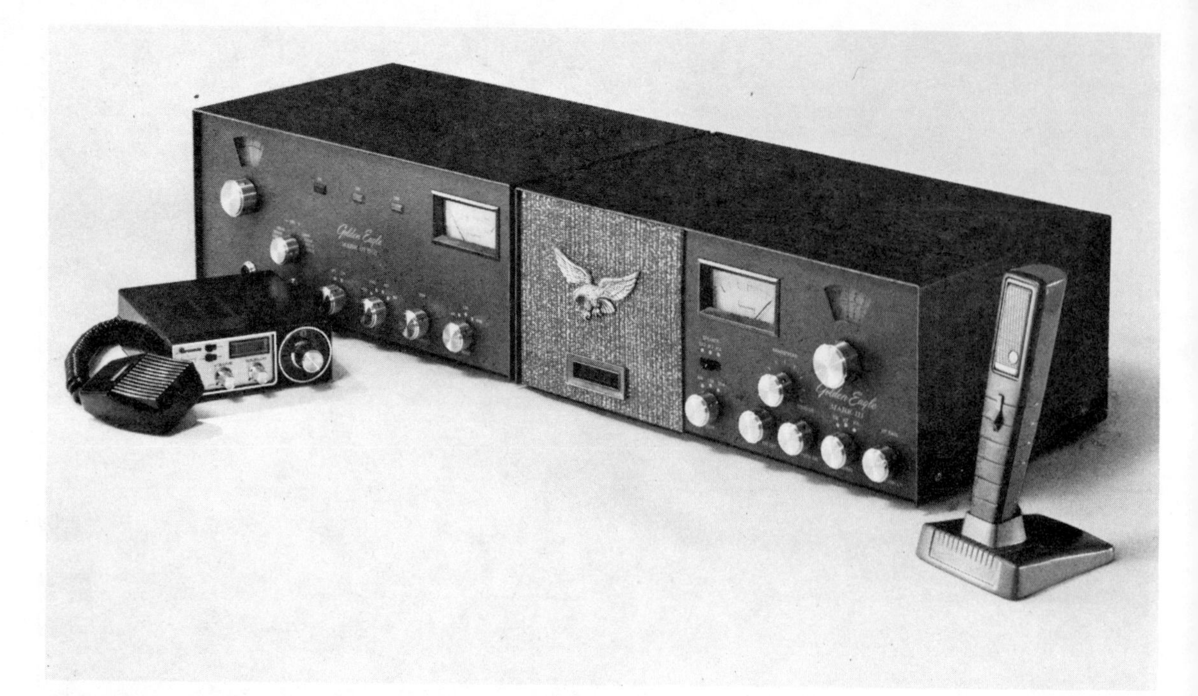

CB's Golden Oldie

Considered by people who don't know her to be an obsolete relic of the tube age, the Browning Golden Eagle Mark III is actually one of the top performers around.

The Eagle consists of two huge units, a separate AM/SSB transmitter and receiver, each with separate power supplies. Tuning is done by twirling a dial instead of clicking a knob, there are no transistors or integrated circuits, you have to let the units warm up before you can communicate, you can't fit an Eagle in the car or in a desk drawer, and there's point-to-point hand wiring instead of printed circuits. It would seem to be hopelessly behind the times, yet people are often willing to wait a year to get one.

Why? Maybe it's nostalgia for the good old days. Maybe people connect size and weight with power. But mostly it's because the Eagle gives its user such a feeling of being in radio! *You don't select a channel, you tune in a frequency. If something goes wrong, you don't ship it back to its maker, you flip open the top and trace the wires and check the tubes. If you think you are a better engineer than the people at Browning, you can try different tubes and fiddle with a dozen knobs to extract the last tiny degree of improved performance. You don't sell it at a swap meet when it's three or four years old, you send it back to Browning for a reconditioning on its 10th birthday. You don't own it, you live with it. It's an entirely different relationship than is possible with any regular CB.*

The unit has a distinctive "clank" that everyone hears when you go on the air, instantly identifying you as an owner of the rare breed. Sound quality and range are tops, and will be the envy of anyone with even the most modern transceivers.

Browning is an old, family-owned operation, up on the shores of Lake Winnipesaukee in Laconia, New Hampshire. It's the second oldest radio manufacturer in the country, going back to the early 1920s when Glen Browning pioneered some of the most important radio advances. The company has been in CB since the start of Class D in 1958. Production is very limited, but quality control very, very good, and all the company's products, which include smaller one-piece mobile units as well as the Eagle, are well regarded in the field. The Eagle transmitter sells for $410, the receiver for $285.

How Far Can I Talk?

The first question asked by prospective CBers and the most common topic among veterans is "How far can I talk?" There is no definite answer, but we'll try to give you some guidelines.

In a big city, crowded with tall buildings, car-to-car communications range seldom exceeds 5 miles, and is often as little as a few blocks. On highways near a big city, car-to-car distance is usually 10–15 miles. In the country, car-to-car distance can hit 20 miles or a bit more. And across flat desert or a body of water, you might go 30 miles. Base station-to-car communications range can be up to twice the above figures, depending on the height and design of the base station antenna. Base station-to-base station range could reach 100 miles or more depending on antenna height and design, and terrain, but is normally within 30 miles.

The higher your antenna, the further your voice will carry. The most notable improvement comes when you fasten your "stick" above the tree tops, because vegetation absorbs radio signals. As a rule of thumb, if your antenna tip is 20 ft above the ground and carries your signal 7 miles: raising the antenna to 40 ft will increase your range to 11 miles; to 60 ft, 14 miles; 80 ft, 20 miles; to 200 ft, 30 miles.

An antenna with "gain" can multiply the effective power of your transmitter. An antenna with 3 dB of gain effectively doubles the power, and increases your range by 4 or 5 miles, a unit with 6 dB gain multiplies your effective power by 4, and increases your range by 10–15 miles, an antenna with 10 dB gain increase your effective power 10 times, and your range approximately 35 miles.

Any increase in range caused by using higher antennas has to be weighed against signal loss caused by having to use longer antenna cable. RG8/U cable has a loss of 1 dB, or 20% per 100 ft. RG58/U cable costs about half as much as RG8/U and is much thinner, but it loses about 2.6 dB or 45% per 100 ft, and any dollar saving must be calculated against the need for a more expensive antenna to make up for the cable loss.

Reach for the Sky: All You Need To Know about Antennas

Antenna manufacturers like to say that the antenna is the most important part of a CB radio system. We won't go that far, but we will say that the antenna is the most easily neglected part of the system. That's because most antennas are fairly dull assemblages of aluminum tubing that are no match for the visual appeal of a transceiver laden with flashing lights and swinging meter needles. Don't try to economize on an antenna to afford a more expensive transceiver. Transceivers are all pretty much the same; they all put out about the same 3½–4 watts of power. Antennas vary tremendously. A poor one can waste 3 of those precious 4 watts. A good one can increase effective power 100 times.

Antennas are not made just any old length. Their size is carefully calculated so they will respond ("resonate") on a particular wavelength and frequency, and if you try to use an antenna designed for one frequency on any other frequency, you will lose signal power and possibly damage the transmitter. This means that the AM antenna on your car won't do a very good job on CB. It also means that if your CB antenna is tuned for best power on Channel 20, it will be less efficient on Channel 10, and still less efficient on Channel 2.

The length of a radio wave pushed out by a CB transmitter operating in the 27 MHz frequency range is 11 meters, or roughly 36 feet, and, theoretically, an antenna should be that long. However, no car, and few house roofs, could handle such a piece of pipe, so compromises are made. About the biggest base antennas you will see are ½ wave length, or 18 ft from top to bottom, but ¼ wave length (9 ft) units are more common.

The first CB mobile antennas were 9-ft steel whips. For optimum performance and equal signal strength in all directions, the antenna should be mounted in the center of the car roof. If you ever put a 9-ft antenna, however, on top of a 5-ft car, you'd have difficulty with garages, trees, and signs. So, to make things more convenient, antenna makers developed the "loaded" antenna, where some of the vertical length of the antenna is replaced by a coil of wire. Most mobile antennas use these loading coils, and measure somewhere between 20 to 50 inches in height. Shorter antennas are more convenient, taller ones talk further.

Remember that the basic antenna is ½ wavelength. When a ¼ wavelength antenna is used, the missing ¼ is provided by a reflective surface or "ground plane." In base antennas, the ground plane takes the form of three or four horizontal or downward tilted "radials" mounted at the bottom of the vertical shaft. On mobile antennas, the ground plane is the car body itself.

If the antenna is not in the center of the ground plane, the signal radiated will not be equally strong in each direction and reception range will be similarly unbalanced. An antenna mounted in the center of the roof gives the most even pattern, and as you approach the corners of the car, the pattern distorts more and more. The long whip mounted on the left side of the rear bumper may look like "Highway Patrol," but it works badly, sending most of its signal to the right front, and bending almost horizontal at high speed.

Mounting an antenna in the center of the car roof is problematic. Any hole you drill can leak if you don't mount the antenna properly, or if the antenna gets jarred by an overhead obstruction. Also snaking an antenna cable through the headliner of the car is a most unpleasant task. An easier way to mount an antenna is to use one of the new magnetic mounts such as the Turner SK-910, or Hy-Gain Hellcat 3 which will stay put at several times the legal speed, need no drilling, and can be easily hidden when you park your car in a high-crime neighborhood.

If you don't want to have wire running across your roof and in the car window, you will do well with one of the very popular

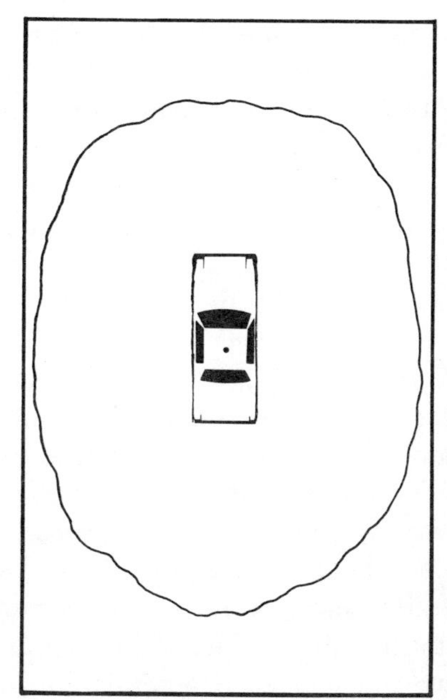

An antenna mounted in the center of the car roof provides the most even radiation pattern.

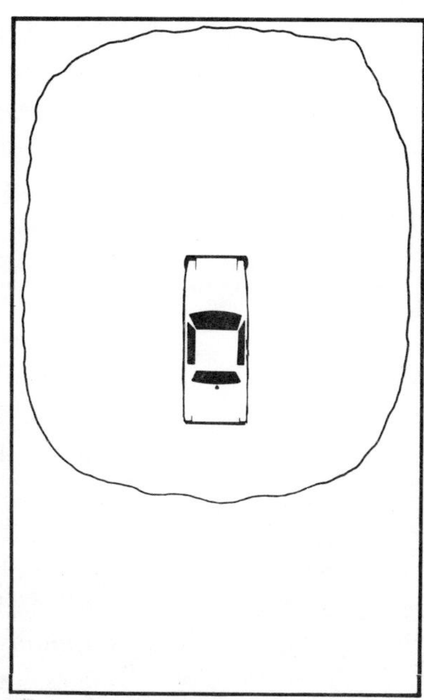

An antenna mounted on the front edge of the trunk has a fairly even radiation pattern, with a bias toward the front.

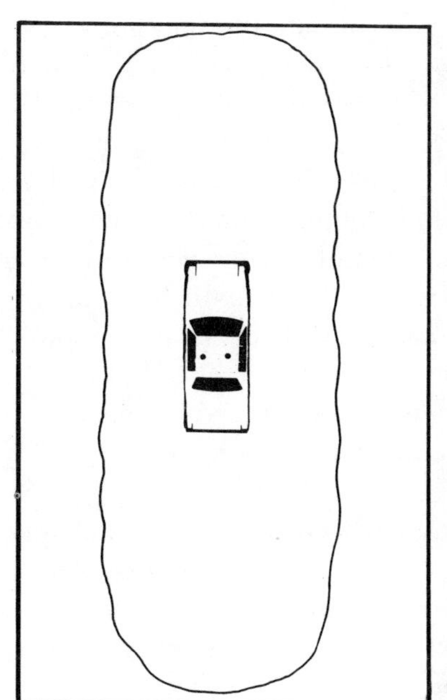

Cophased side-mounted antennas provide gain up and down the road, with a consequent loss to the sides.

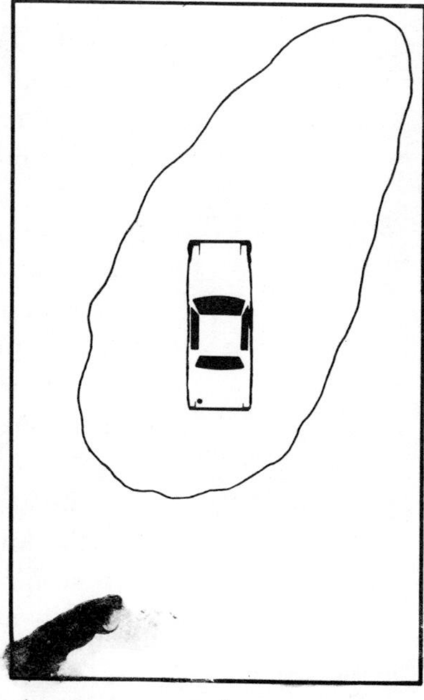

An antenna mounted on the left rear bumper concentrates power in the opposite direction, toward the right front.

trunk lip mounts such as the Turner SK-211, Antenna Specialists M-125, or Hy-Gain Hellcat 4. These units mount without any drilling (which could lower the trade-in value of your car), but they flip forward when the trunk is raised, and can scratch the roof, particularly if you have a small car. There is another type of antenna mount that goes in the trunk groove, and holds the antenna vertical when the trunk is raised. "Groove mounts" require two small holes in the groove area, but no damage is done to the outer body. Trunk mounts of both types provide a fairly even radiation pattern, with a slight bias to the front, which is fine for cars, as you will generally have more interest in what lies ahead than what you've already passed.

Most 18-wheelers use two antennas, one mounted on each rear view mirror, in what is called a *cophased* array. This arrangement is important because the signal does not get blocked by the shifting trailer and it works nicely because the distance from one antenna to the other is just about ¼ wave length or 9 ft. A lot of CBers have put cophased antennas on their 4-wheelers recently and most antenna manufacturers make cophased models for cars, but they are really more for visual effect than for effective communication. On a car, a cophasing array will give you a marginal gain (maybe 10% more range) to the front and rear, with a similar *loss* to the sides.

Cophased car antennas are available for groove mounting on both sides of the trunk, both sides of the hood, drilled into the roof, or clamped onto the rain gutters. The clamp-on units, either in the cophased or conventional single antenna design, require minimal installation time, and are easily removed for security or transfer to another car. They are very good for people who frequently rent cars and can't be without CB.

Most manufacturers make a basic antenna and sell it with a variety of mounts and mounting accessories. The spring is mandatory equipment, as it lets the whip bend out of the way of obstructions, instead of smashing into them, damaging the antenna and the car. Swivel balls are necessary if the antenna is mounted on the side of a vehicle, or on a slanted surface, such as a fast-back or hatch-back roof.

Loading coils come in various shapes and sizes and appear in different places on the antenna. The bigger coils have heavier wire inside and are less likely to heat up and waste power. The coil is actually used to radiate and receive signals, so the higher up on the antenna it is, the better it works. A top-loaded coil is best in theory, but they have a habit of smashing into things, or catching on obstructions and ripping out the whole antenna. And because they are bulky and top heavy, they sway a lot when the car is moving, which puts them out of tune (and nearly off the air) as they swing low at high speeds. Center-loaded antennas such as the Super Hustler, are made rigid enough to avoid swaying, and stand quite tall for good reception, but for the vast majority of CB units, a base-loaded whip works just fine, looks good, and won't malfunction so easily.

A number of manufacturers, such as Shakespeare, which specializes in such products, make fiberglass antennas that have the coil wound around the entire antenna shaft from top to bottom for even signal distribution. Glass antennas are supposed to be more flexible than steel, but when they can't bend, they snap.

Special Cases

If you drive a van, you may love the look of cophased antennas on each mirror, or a 9-ft whip mounted on the side, but neither arrangement is as good as a single base-loaded antenna in the center of the roof.

If your transportation is a fiberglass Corvette or Saab Sonnet, you have no ground plane. A conventional antenna won't work unless you attach copper screening or large amounts of metallic tape to the underside of the trunk or roof where the antenna is mounted. An alternative would be to use a marine antenna such as the 8-ft tall Antenna Specialists M-223 or 4-ft tall ASMR94, both of which work without ground planes, and are fine for fiberglass cars, or motorcycles, for that matter.

Antenna Specialists, by the way, is probably the largest manufacturer of communication antennas in the world, and has hardware to accommodate any vehicle, conventional or offbeat.

Not only can they sell you a shaft for your Chevy, but they've got antennas for mobile homes and jeeps, an antenna (M-306) that works on snowmobiles, and even a bicycle whip (M-307) that not only keeps you in touch, but carries a warning flag to keep 4-and 18-wheelers from touching your 2-wheeler.

Up on the Roof

Antennas designed for base use come in even more variations than the mobile units, and because they can be bigger, they can do a lot more.

The two basic classes of base antennas are *omnidirectional*, and *directional*. Omnidirectional antennas transmit and receive with equal ability in all directions. Directional antennas, also called "beams," work in a much narrower area, and because they concentrate their power, can talk farther than omnidirectional types.

The basic base station antenna is the 1/4 wave-length ground plane, such as the Antenna Specialists M181, Radio Shack 21-901 or Lafayette 42 R 01562W. These types provide the home equivalent of the 9-ft mobile whip antenna. Instead of using the car body as a ground plane, the base antenna has three or four "radials" attached to the antenna at the bottom of the vertical shaft.

Omnidirectional antennas usually look like this, with a vertical whip radiator, plus three or four horizontal radials to reflect the radio waves, taking the place of the earth's natural ground.

A beam antenna uses three or more elements. One actually generates the radio waves, a slightly shorter one directs the waves into a specific point, and a longer "reflector" element cuts down on radiation in the opposite direction by bouncing the signal forward.

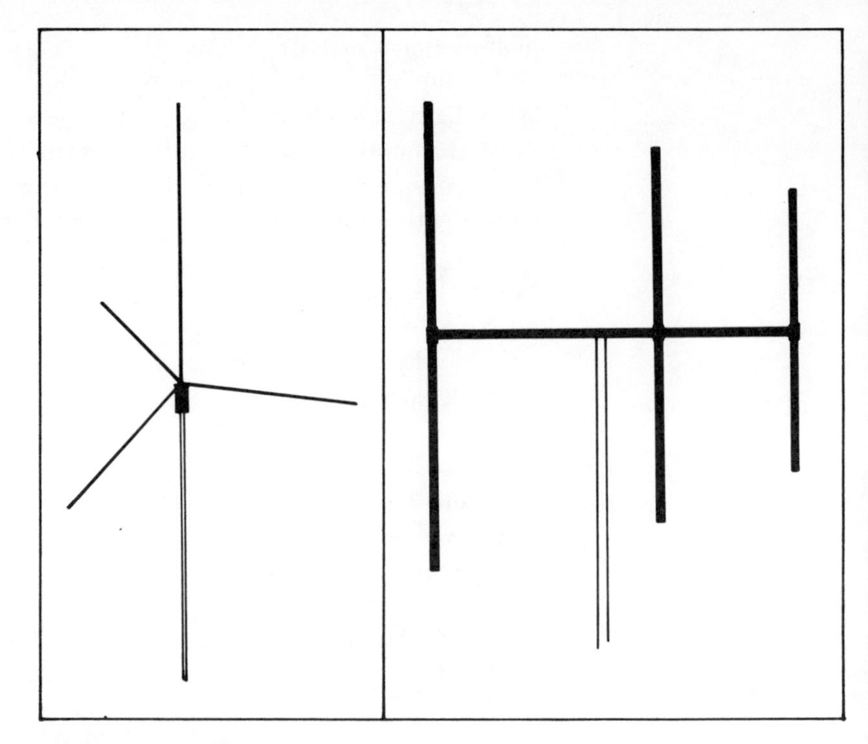

500 Legal Watts!

While the FCC limits the output of CB transmitters to 4 watts, it is possible to build an antenna that legally increases the effective power. Most of these units are directional beams, which essentially take the power that is normally wasted by being radiated in all directions, and saves it by pushing it only in the desired directions. The simplest beam has three elements, like the Antenna Specialists M-202 which has a gain of 9.75 dB for an equivalent power of nearly 40 watts. More complex antennas can have dozens of elements, and one of the biggest, the Wilson Super Laser 500, is 40-ft long and 18-ft tall and has a gain of 18 dB, giving you an effective radiated power of about 500 watts!

If you decide to use one of these big-beam antennas, you will need something stronger than a piece of TV antenna mast to mount it on. Rigid towers, similar to the ones used for professional broadcasting stations, are available for CB use by such manufacturers as Rohn and Antenna Specialists. There are some units that tilt or crank down for servicing so you don't risk your neck climbing up to the top. Don't forget the legal-height limits.

Multiple Antennas

Many people who have beam antennas also have an omnidirectional antenna. The procedure is to listen with the omni until you hear someone you want to talk to, and then rotate the beam toward the target and switch it on for the conversation. The strongest rotator is made by Cornell-Dubilier. Antenna switches are all pretty much the same, and are offered by most CB accessory makers. You can save money on a two-antenna installation by using an antenna switching relay up on the roof or tower, so it will only be necessary to have one cable running down to the transceiver.

Polarity

Radio waves are "polarized" either horizontally or vertically, moving either parallel to the earth's surface, or perpendicular to it. TV uses horizontal polarization, which is why TV roof-top antennas are horizontal. Since it is not practical to use a horizontal antenna on a moving vehicle, CB uses vertical polarization, with antenna shafts aimed at the sky.

However, when CB is used to cover long distances (up to the 150-mile FCC limit), a signal that started out vertical may be twisted as it bounces off the ionosphere and arrives at the receiving antenna in the horizontal form. To take care of this problem, there are beam antennas available from most manufacturers that work in both horizontal and vertical modes. Two base stations can communicate effectively with horizontal polarization, eliminating the noise coming from conventional vertical transmission.

The Numbers

Antennas don't have too many specifications other than their physical dimensions. The few electronic numbers you will see are easy to understand.

Gain refers to the antenna's "amplifying" ability, or advantage over a theoretical "isotropic antenna." Gain is expressed in dB but can be converted to wattage increases. As a rough guide, 3 dB gain doubles the effective radiated power; 6 dB multiplies it by 4; 10 dB multiplies it by 10; 20 dB by 100.

Front-to-back ratio refers to the antenna's ability to reject a signal coming from the back when you want to hear someone to the front. This is expressed in dB, usually in the 10 – 30 dB range, with bigger numbers better.

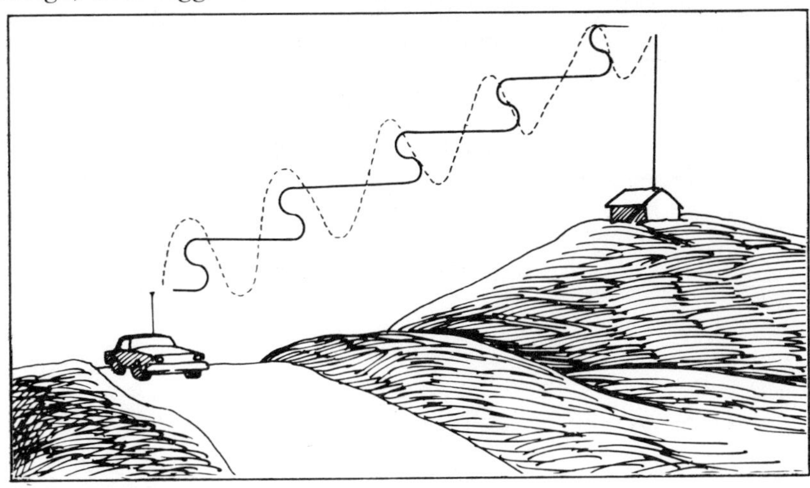

With the normal vertically polarized radio waves used in CB radio, the electrical field moves perpendicular to the earth's surface, while the magnetic field goes parallel to the earth.

SWR, or standing wave ratio, is the ratio of maximum to minimum voltage at various places along the antenna cable. A high ratio indicates that a large percent of the power pushed out of the transmitter is not going out the antenna, but is bouncing back through the cable and being wasted. SWR is a function of many factors, including the weather and proximity to other objects, as well as antenna design. A manufacturer's rating should be used as only a rough guide to the best attainable figure. A ratio of 1:1 is the theoretical best, but anything under 1.5:1 is considered acceptable.

Power handling capacity indicates the number of watts the antenna can handle without being destroyed. These figures are often 100 times the legal 4 watts and to some degree are just advertising claims. But the figures do also indicate the antenna's ability to dissipate power that could otherwise generate heat instead of signal, particularly when a loading coil is used.

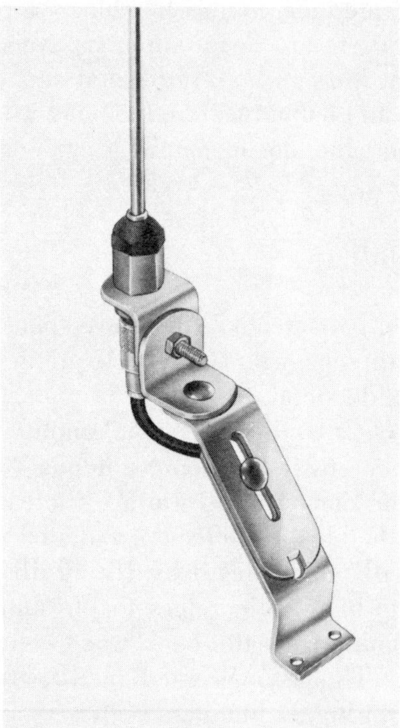

Trunk groove antenna mounts attach with screws mounted in two small holes drilled in the channel that goes around the trunk under the lid. These antennas don't tilt forward, and are a good bet for small cars.

Trunk lip antenna mounts clamp on the front edge of the trunk lid and do not require any holes to be drilled. They tilt forward when the trunk opens, so watch out for scratches on the roof.

Gutter-mount antennas clip onto the rain gutter over the car windows. They get the antenna up high without drilling through the roof, and are easy to remove if you park in a high-crime area or switch cars.

The Turner SK 910 uses a magnetic base to stick to the center of your roof for even radiation in all directions, yet eliminates drilling through the roof and snaking wires through the headliner. And you can take it off when you park. Suggested retail price is $28.

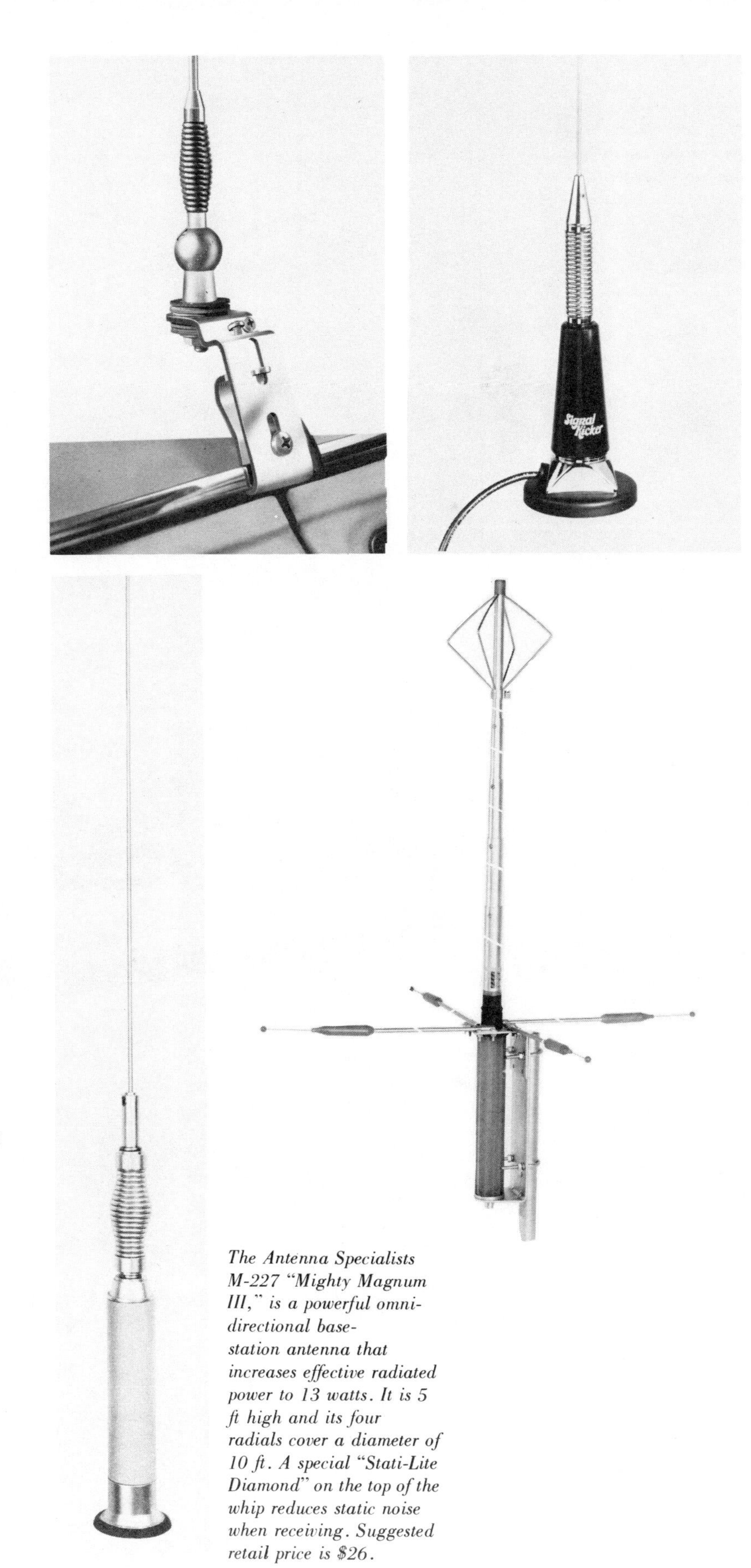

The Antenna Specialists M-125 is a classic mobile antenna. It uses a base loading coil to reduce overall height to 46 inches, making it practical for roof-top installations, but it can also be mounted on the cowl, trunk, or fender. It comes with a 17-foot coaxial cable and connector. Suggested retail price is $23.50.

The Antenna Specialists M-227 "Mighty Magnum III," is a powerful omni-directional base-station antenna that increases effective radiated power to 13 watts. It is 5 ft high and its four radials cover a diameter of 10 ft. A special "Stati-Lite Diamond" on the top of the whip reduces static noise when receiving. Suggested retail price is $26.

The Wilson Super Laser 500 is the biggest CB antenna around (40 ft long, 18 ft high). It's not for everyone, but if you want to travel far, this is your ticket. It has a gain of 18 dB, for an effective radiated power of about 500 watts! It works with both horizontal and vertical polarization, and is designed to survive winds of over 100 mph. Weight is 70 pounds, so make sure you use a very strong tower. Suggested retail price is $440.

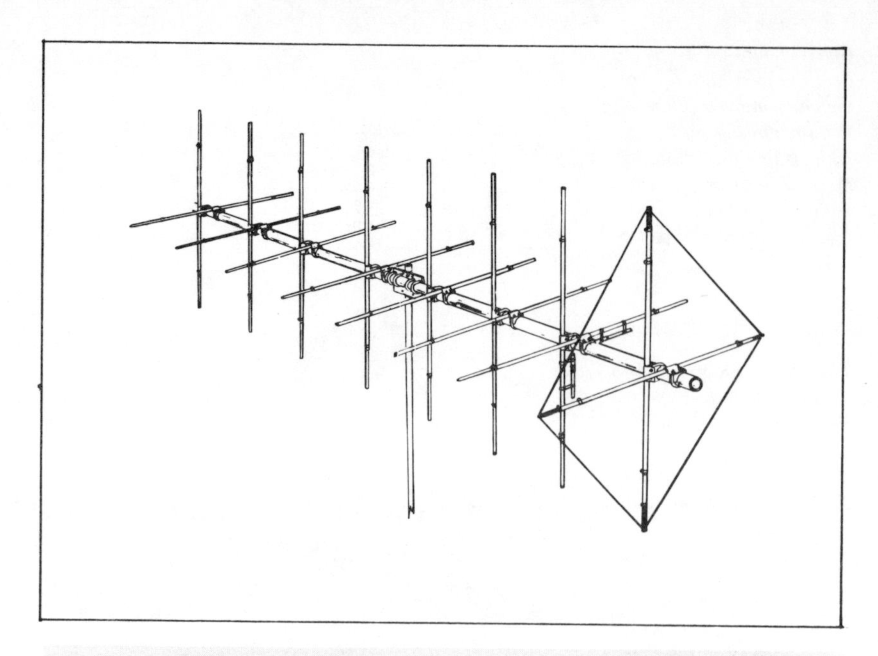

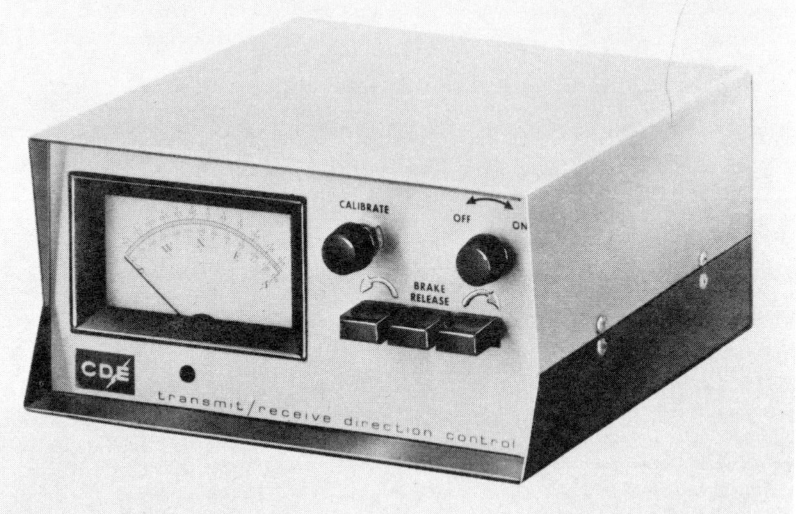

The strongest antenna rotator is the Cornell Dubilier "Ham-II." It can handle the heaviest antennas, in the strongest winds, and has a gentle braking system to minimize stress on the rotator, antenna, and mast. The control unit which sits near your transceiver has a meter calibrated in compass directions. Suggested retail price is $160.

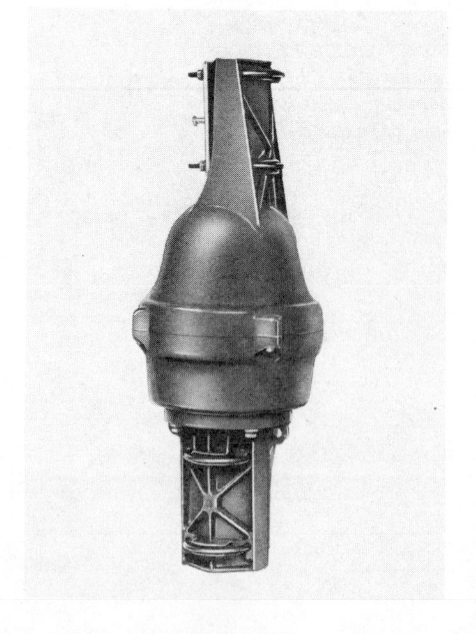

Putting Ears on Your Wheels

While anyone who can read and knows how to use a screwdriver can install a CB set, many people are really afraid of any technical chore. If you're like that, a professional installation is your best bet. Many CB dealers will put the set in for you, and while the day of the free installation is just about over, you should be able to get a good price ($10–15), if you have the unit installed where you bought it. If you go to an installation specialist, expect to pay about $35 for a transceiver and "trunk lip" antenna. For more complex antennas that involve drilling holes in the roof or fender and snaking cable through the car's headliner, the price could double.

You should be able to get a CB installed at most car stereo centers and major auto accessory centers if your CB dealer can't do the job. But there's a big difference between installing a CB and installing it right, so try to find someone who knows what he's doing. Check with other CBers for endorsements and ask the prospective installer if he has an "SWR meter." If you get a blank stare instead of a yes, go somewhere else.

Doing It Yourself

No book can possibly explain how to install every type of antenna and CB transceiver, but if you combine the manufacturer's instructions with the information here, you should be able to get on the air quickly and easily, and save $30 or more in the bargain.

Antennas come in an awesome variety, with many variations in mounting procedure. What we will discuss here is the most popular type, and the one we recommend—the trunk lip mount which gets the antenna near the center of the car for an even radiation pattern, and doesn't require any drilling through the car body that can cause leaks and will lower the car's value at trade-in time. We did our test installation with the Turner SK211 Signal Kicker, which is typical of the trunk-lip type and a very fine antenna.

Unwrap the antenna and spread out the parts in a convenient location where you're not likely to lose anything. Take the antenna base and loosen the two set screws using the hex-head "Allen" wrench provided. Raise the trunk and position the base on the back edge of the trunk, using the protective pad the manufacturer provided. Have someone hold the base in position while you tighten the screw from below. Make sure the screws are tight; they hold the antenna on the car and also provide the vital ground connection.

Then remove both top and bottom of the back seat from the car. These can usually be removed with a simple push or tug in the right direction. If you can't figure it out yourself, ask a friendly mechanic or your car dealer. Using a Phillips screw driver, poke a hole through the cardboard panel behind the seat, low, and on the left side. Snake the wire from the antenna to and through the hole. Secure the cable every foot or so to the top of the trunk compartment with heavy tape or plastic wire clips, such as Radio Shack's # 278-1635.

Using a Phillips screwdriver, remove the screws that hold down the metal "threshold plate" (under the doors) on the left side of the car. Pull out any nails or screws that hold the carpeting in place, and push the carpeting back from the door sill. Run the antenna cable from the trunk hole, under the back seat, and then under the carpeting until it comes up under the dashboard. Replace the carpeting and the threshold plate(s).

Find a convenient place to mount the transceiver. This will probably be under the dashboard, but you might want to consider hiding it in the glove compartment or mounting it on the transmission hump with a security mount such as the Shur-Lock CB-700HM. Make sure the CB will not interfere with the operation of any car controls, and that it's easy to reach with seat and shoulder belts fastened. If you mount the CB far "up" under the dash it will less likely be seen by thieves, but it will more difficult and less safe for you to use. If the CB is to be operated by the passenger as well as the driver, keep the unit in the middle of the car so the mike cord won't have to be stretched in front of the driver when the passenger is transmitting.

Using the transceiver bracket's screw holes as a pattern, drill two holes in the bottom of the dashboard, and fasten the bracket using the screws, washers, and nuts provided.

Attach a connector to the antenna cable. Many antennas, such as the Turner, come with connectors and instructions. If your antenna did not come with a connector, you will need to buy one at an electronic parts store. It's called a PL-259 and sells for under a dollar. You will also need an RG58/U cable adaptor collar, which sells for about fifty cents. Separate the PL-259 into its two pieces, and push the antenna cable through the adaper collar and the outer shell of the PL-259. Using a razor blade, slice a shallow ring around the cable ¾ of an inch from the end and make another cut from the ring to the end of the wire. Peel off the insulation. Slide the adapter collar down to the end of the insulation and peel the braided outer shield wire back over the collar. Trim to ⅜ of an inch. Carefully, using either the razor blade or a wire stripper, remove a ⅝ of an inch of insulation from the center conductor of the cable, and push it through the hollow pin on the remaining part of the PL-259. Screw the adapter collar into the PL-259. Solder the center conductor to the tip of the hollow pin. Solder the shield to the plug through the holes. Screw the PL-259 shell over the body. Don't insert the PL-259 into the antenna socket on the CB set yet.

There will be two wires coming from the back of the transceiver, usually a red one containing a fuse socket, which goes to a source of positive electricity, and a black wire which goes to negative source or "ground." The positive wire can be connected to any number of places, the ignition switch and the fuse block being most often suggested. It is usually a lot easier to tap into the "hot lead" that supplies power to your regular radio or the

Assemble the antenna base and hook it around the trunk lip

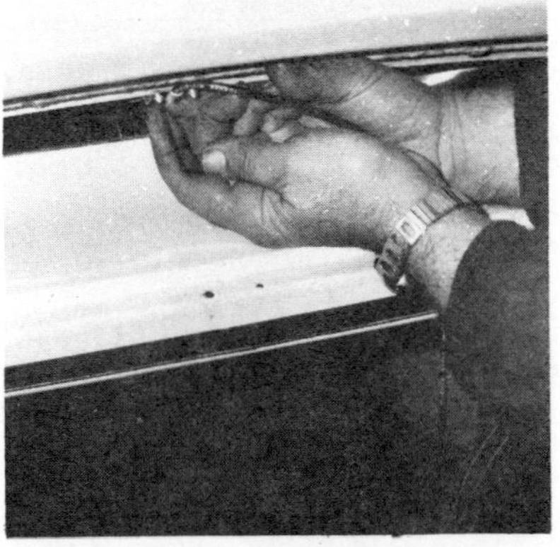

Secure the base to the trunk lip by tightening the Allen screws.

Remove the rear seat, and poke a hole through the trunk wall, to pass the antenna cable.

Remove the threshold plate from under the door, and pull the cable from the trunk.

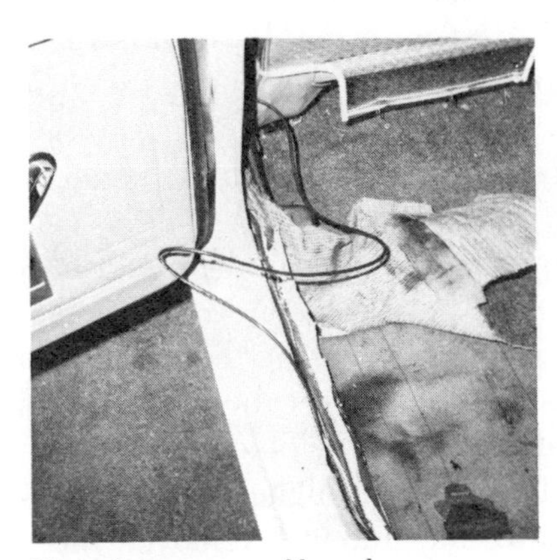

Place the antenna cable under the carpeting, and run it up to the front of the car.

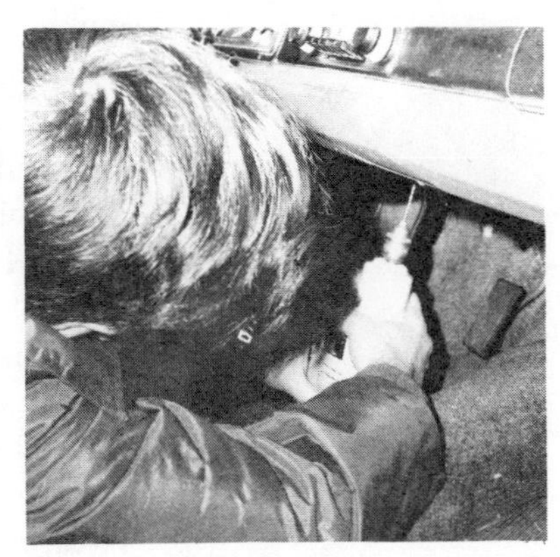

Using the transceiver bracket as a pattern, drill holes through the bottom of the dashboard.

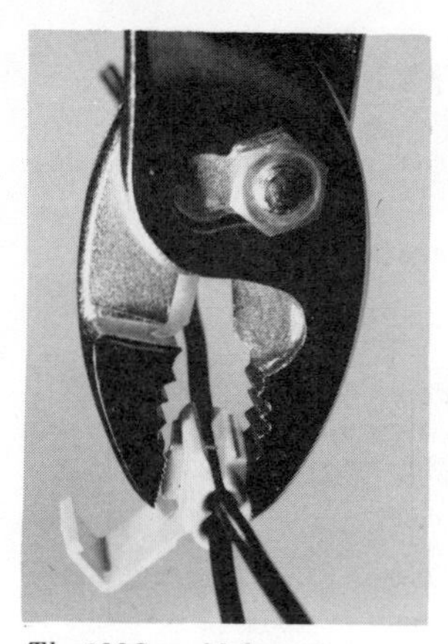

The 3M Scotchlok connector makes it easy and safe to tap into the wire that feeds your AM radio or cigarette lighter to power the CB.

cigarette lighter. As long as you don't light up while you are transmitting or play both radios at the same time there will be no danger of overloading any circuit. The best way to tap into the wire (a mechanically-inclined friend should be able to point it out if you can't find the hotwire yourself) is with a 3-M "Scotchlok" connector, available from most electronics supply houses and many hardware stores. With this connector you won't have to bother with a soldering gun or tape. The connection will be mechanically and electrically secure and perfectly insulated, yet it can be disconnected easily whenever necessary. The ground wire can go to almost any screw you see going into the metal body of the car.

Mount the CB into the bracket with the screws provided, and tuck the antenna cable and power and ground wires up out of the way, securing them with cable clamps or wire ties for safety and a professional appearance. Mount the microphone bracket in a suitable place, where you can get to the mike and where the mike cord won't interfere with your driving. If the trade-in value of your car is an important consideration, mount the bracket below the dash where the screw holes won't show. There are mike brackets with adhesive backs that can be stuck to a smooth surface, but most dashboards have a leather or wood grain that won't stick too well. And if you use a magnetic mike mount , cover the magnet with soft tape to prevent scratches.

Buy or borrow an SWR meter and connect it between the antenna cable and the CB set. Turn the SWR-meter switch to the forward ("FWD" on most units) position, turn on the CB set and hold down the microphone button so you will be transmitting (make sure there's on one else on the channel). Adjust the SWR-meter calibration control until the needle hits the indicator line on the right end of the meter scale. Switch from the forward to reflected ("REF") mode, and still transmitting, make a note of the meter reading. If you expect to be using all CB channels, make this test on a middle channel such as Channel 12. If you want maximum power for emergency use, do the test on Channel 9 (but be quick, so you don't tie up the emergency frequency). If you only expect to use one channel, make the test on that channel.

While you are watching the meter needle and holding down the mike button, have someone raise or lower the antenna a fraction of an inch at a time, using the Allen wrench that comes with most antennas. When you get the lowest reading, hopefully about 1.1, tighten the antenna in place. The antenna should be shorter for best performance on higher channels. Make sure the person who is adjusting the antenna removes the Allen wrench and walks away from the car while you make each measurement.

Disconnect the SWR meter and its cable. Connect the antenna cable to the transceiver. You're now ready to go on the air.

Domestic Ears: Installing a CB Base Station

Most CB dealers and communications equipment specialists charge $100 or more to install a home antenna and transceiver, so the savings can be quite considerable if you do it yourself. And almost anyone can. Base antennas vary greatly in design, but the installation is standard; put the antenna on a pole, and put the pole on the roof. Or put the pole up first, and then add the antenna.

The higher you can get your antenna, the farther out you will be able to talk. FCC regulations limit antenna height to 20 ft for directional "beam" antennas, but the limit for omnidirectional antennas has recently been raised to 60 ft. These limits are normally considered to apply to overall height above your roof, but if you happen to live in a 400-ft tall building or near a 200 ft hill, there is nothing to prevent you from going 20 ft above them, providing you use an antenna with enough gain to make up for the signal lost traveling through such a long antenna cable.

For fairly light antennas (up to, say 30 lb, including any rotators) you can use any standard TV-type antenna mast. These are available in 5 and 10-ft lengths, and telescope together to make longer masts.The mast can be mounted in a variety of ways. There are base plates designed for flat roofs and peaked roofs, and there are brackets designed to attach to chimneys, pipes, and walls. Whatever method you use, make sure everything is secure and sturdy, because a big antenna can catch enough wind to topple a chimney! Use guy wires every ten feet if

A full-size 1/4 wavelength whip antenna is 9-ft tall, too high for most garages. A gutter clip, such as the Shakespeare 4107 shown here, lets you tuck the antenna down out of the way to fit into close quarters.

For dependable mobile communications beyond a few miles, both car and base should be equipped with preamplifiers, such as this Telco BoosTwenty. It increases receiver sensitivity by a stated 20 dB, for a big increase in range. The unit is rugged, and can be mounted in your trunk or behind the dashboard. It works off the car's power supply and can either be left on all the time, connected to a switch, or wired to go on when you turn on the transceiver. Suggested retail price is $35. Telco also makes a preamp for home use, the "Base Booster," priced at $60.

your mast is more than 20 feet tall, and put the guy wire anchors into the roof before you raise the antenna. Metallic guy wires reflect radio waves and can hurt your antenna's performance, so use nylon cable if possible. If you must use metal, cut it into lengths of seven feet or less, splicing the short lengths with insulating "eggs" available at electronic parts stores.

Most base station antennas come in knocked-down form, and can be assembled with a screwdriver and a couple of wrenches. Before you start assembly, unpack the parts and sort them into piles of identical sections, and then count them against the parts in the instructions, and if something is missing, call your dealer.

If you are using a tall mast, it will be necessary to mount the antenna and cable on the mast before going up on the roof. Spray the cable connections with a silicone such as "CRC", and attach the cable to the mast with nylon wire ties every few feet. Make sure you leave plenty of slack in the cable if you are using a beam antenna with a rotator. Attach the guy wire collars and the guy wires at this time.

With help from an appropriate number of friends, "walk" the antenna up the side of your house, the same way you would raise a ladder. Get your crew up on the roof, and pull up the antenna. Walk the antenna up vertical and attach to the base, with your friends holding the guy wires for safety. Then attach the guy wires to the anchors, and tighten the turnbuckles until the antenna is secure. Be sure to use liberal amounts of waterproofing compound where the guy-wire anchors are screwed into the roof and on any screws that secure the mounting base into the roof.

If you are using a short mast, as we did in our test installation with the Radio Shack's excellent 5/8 wavelength antenna (#21-1133), you can put the mast up first, and then attach the antenna. We used a pair of wall mounting brackets just below the peak of the roof, with about 6 ft of mast.

Climb up to the mounting spot, and using a ruler, note the location of the best spots for the backet screws. Climb down, attach the brackets to the mast, the proper distance apart (at least 3 feet for best stability) and then go back up and screw the brackets to the wall. Then bring up the complete antenna, mount it by the mast, and connect the cable.

Attach wire "stand-offs" every 4 ft onto the roof and the side of the building leading down to the entrance point. Make sure the "stand-offs" are not the kind designed for flat TV-antenna wire, but have holes big enough for round coax plus a rotator wire if used. Either run the wire through the window, using a piece of garden hose as an abrasion shield, or drill a hole through the wall, insert a weather resistant wall tube such as Radio Shack's 15-200 and run the cable through that. Drill through the frame. Try to keep cable distances as short as possible to reduce signal losses and try to bring the cable into the house very close to the place where the CB will be used. If it is necesary to travel over 50 ft, use the heavier RG8/U cable, and if you have to run the cable

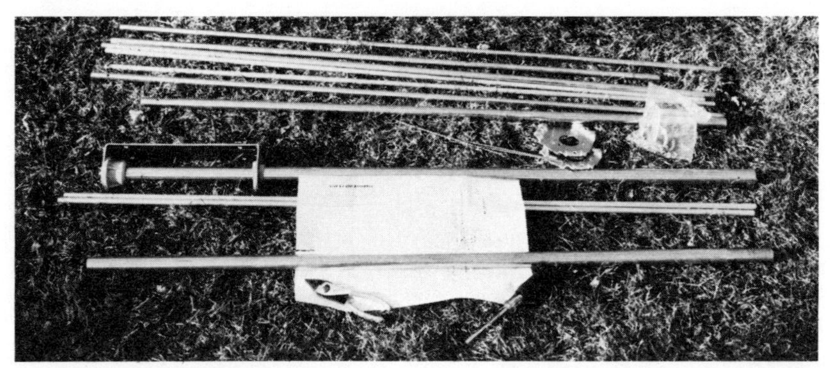

Spread out the antenna sections and check the parts list to make sure you have everything.

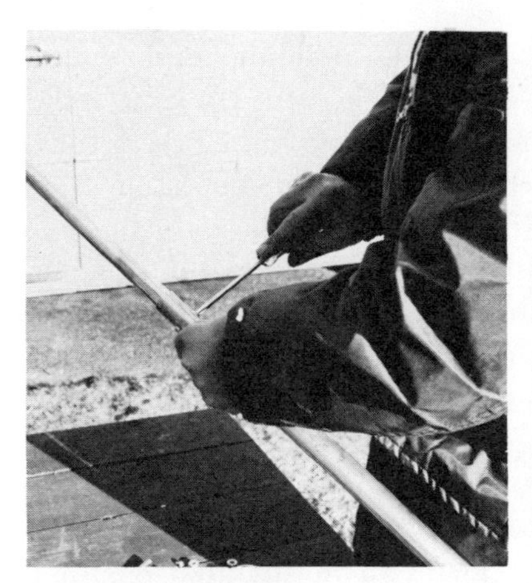

Carefully assemble the antenna, following the manufacturer's instructions, one-at-a-time.

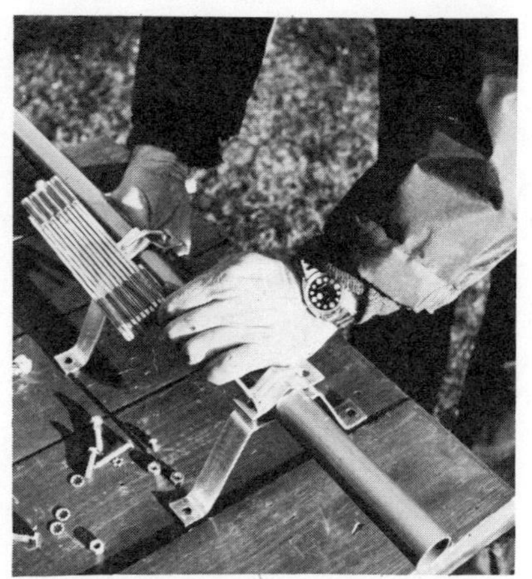

Attach the mounting brackets to the antenna mast, positioned for a secure mounting.

Attach the mast to the wall and mount the antenna on the mast. Run the cable down and into the house.

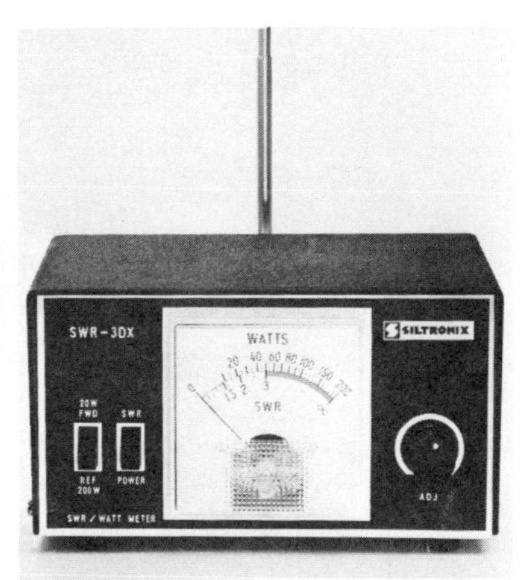

Use an SWR meter, and adjust the antenna length or use a "match box" for proper tuning.

through the walls, try to stay away from power lines and plumbing, and plaster up any holes you've drilled. Put the CB set where it won't be in direct sunlight, or close to a radiator or heating duct.

Attach the connector to the antenna cable. Many antennas come with connectors and instructions. If your antenna did not come with a connector, you will need to buy one at an electronic-parts store. It's called a PL-259 and sells for under a dollar. If you have the thinner RG58/U cable, you will also need an RG58/U adapter collar, which sells for about fifty cents. Separate the PL-259 into its two pieces, and push the antenna cable through the adapter collar and the outer shell of the PL-259. Using a razor blade, slice a shallow ring around the cable ¾ of an inch from the end and make another cut from the ring to the end of the wire. Peel off the insulation. Slide the adapter collar down to the end of the insulation and peel the braided outer shield wire back over the collar. Trim to ⅜ of an inch. Carefully, using either the razor blace or a wire stripper, remove ⅝ of an inch of insulation from the center conductor of the cable, and push it through the hollow pin on the remaining part of the PL-259. Screw the adapter collar into the PL-259. Solder the center conductor to the tip of the hollow pin. Solder the shield to the plug through the holes. Screw the PL-259 shell over the body. Don't insert the PL-259 into the antenna socket on the CB set yet.

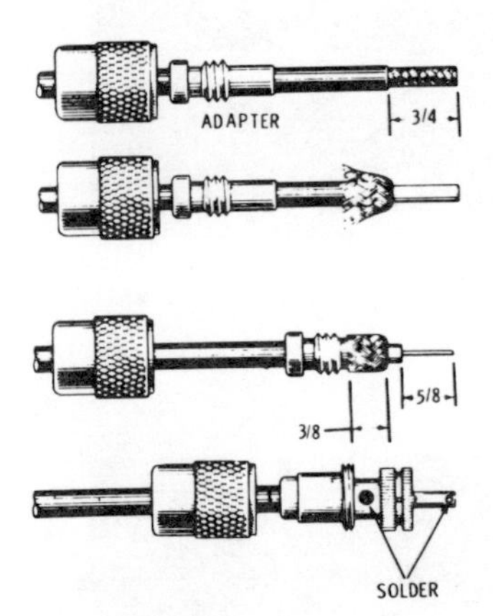

Attach the PL-259 connector to the antenna cable, using a razor blade, wire stripper, soldering iron, and a steady hand.

Buy or borrow an SWR meter and connect it between the antenna cable and the CB set. Turn the SWR-meter switch to the forward ("FWD" on most units) position, turn on the CB set and hold down the microphone button so you will be transmitting (make sure there's no one else on the channel). Adjust the SWR-meter calibration control until the needle hits the indicator line on the right end of the meter scale. Switch from the forward to reflected "REF" mode, and still transmitting, make a note of the meter reading. If you expect to be using all CB channels, make this test on a middle channel such as Channel 12. If you want maximum power for emergency use, do the test on Channel 9 (but be quick, so you don't tie up the emergency frequency). If you only expect to use one channel, make the test on that channel.

If the SWR reading is too high, you can install a "matchbox" such as the Gold Line Antenna Matcher that lets you tune your antenna without the risk of rooftop adjustments.

Disconnect the SWR meter and its cable. Connect the antenna cable to the transceiver. You're now ready to go on the air.

CB Accessories

With CB , like many other hobbies, it's not the initial investment that gets you, it's the upkeep and add-ons. The Electronics Industries Association reports, for example, that the average consumer will spend $150 on his basic CB transceiver, and another $400 on antennas and other accessories (including, of course, a possible second transceiver). There are literally hun-

dreds of CB gadgets and electronic doodads on the market, all of which claim to offer you better reception, stronger transmission, or both. Some actually do deliver on their promises. We'll try to sort them out so you can select the equipment that you really need.

MICROPHONES. The factory mike that comes with a CB is perfectly adequate, but can be improved on. Normal speech has something called "dynamic range," which is the range between the loudest and softest sounds you make. If you speak too softly, you will not sufficiently modulate the carrier wave, and you will not talk very far. If you speak too loudly, you will overmodulate, putting out a distorted signal similar to the sound a cheap radio makes when turned up to full volume. A number of manufacturers, such as Mura, Turner, Shure, and Hy-Gain make amplified, or "power microphones," that increase your modulation to the maximum potential. A power mike has an internal transistor amplifier, powered by a small battery, and an adjustable gain control to regulate amplification. Most of the time you will probably operate at or near maximum. You make adjustments by getting a signal report from someone who hears you on the air, or by using a modulation meter. More sophisticated power mikes include, in addition to the amplifier circuit, a "limiter" which sets a maximum modulation level to prevent distortion. The combination of amplification and limiting is called "compression," because it makes the voice's dynamic range narrower; the softest sounds are less soft, and the loudest, less loud.

Compression is used in recorded music. HiFi purists complain about its unnatural sound, but it's a great blessing in communications. Turner makes a +3 microphone in both base and mobile models with an excellent compressor circuit. It not only provides good intelligibility whether you whisper or shout, but also keeps the volume constant when you move the mike close to or far from your lips.

It's a good idea to always keep your mike close to your mouth when you're mobile so your voice will override noises coming from inside or outside your vehicle. If you drive a particularly noisy diesel or off-road vehicle, a "noise canceling" microphone will help considerably. Made by Turner, Shure, Electro-Voice and others, these mikes were originally designed for industrial, aeronautic, and battlefield applications, and are now finding their way into CB shops.

Noise-canceling microphones have two openings or "ports" that pick up sound. The main port is held close to your lips, and the other opening faces away. Any sound, like wind noise, that is picked up by both ports, is electronically eliminated. Your voice, which comes in only from one port, is allowed to go through. One of the most sophisticated noise-canceling microphones available today is the Telex CB-73 Double Header. In addition to the two ports, it has an internal amplifier for boosting modulation, a switch to defeat the noise-canceling action (if you want to hold the mike further from your lips in a quiet environment), and a unique front hang-up hook that lets you grab the unit in a hurry and speak naturally, instead of going through the thumb-bumping finger dance necessary with most mobile mikes.

Each microphone model is actually a family of related models that work with different kinds of CBs with different kinds of connectors and either relay switching (used mostly on tube-type CBs) or electronic switching (used on most solid state models). Your dealer can make sure you get the correct model. Because of the wide variety of transceiver connectors, most accessory mikes are sold without plugs, which are sold separately by electronics dealers. Putting one on the mike demands a surgeon-steady hand with a soldering iron, and if you are not experienced in such things, have your dealer take care of it.

HEADPHONES, HEADSETS, AND HANDSETS. If your engine is noisy, or if you drive with the top down or the window open, or if your traveling companion hates CB chatter, you need a way to get the CB sounds directly into your ear. Virtually every transceiver has an earphone jack in the back, and any electronics store can sell you an earplug for about $2. You can get considerably better sound by using a hi-fi-style headphone. Recently, companies like Superex and Telex have begun marketing headphones tailored to CB needs. They are lightweight for nonfatiguing long-time use, and their frequency response is designed for maximum voice and minimum noise.

Some cities and states have laws prohibiting wearing headphones while driving, so check your local bears' den before buying a CB headset. And get one of the one-ear-only types, that still let you hear horns and sirens from the outside. To make your driving life even easier, Telex, Superex, and Lafayette have telephone-operator-style headsets that come with boom mounted microphones. These models have all the advantages of regular headphones, but also keep the mike in the right place for best modulation, and keep your hands free from mike holding for more important driving jobs.

Most of these units come with remote switches that you can clip on to your shift lever or any other convenient spot; and there are models, such as the new Superex 10-25VX, that comes with a VOX circuit that automatically switches from receive to transmit when you start talking. VOXes are very convenient and an important safety advantage, but unless they are adjusted properly, they can be triggered by outside noises such as horns, sirens, or thunder. They're best for quiet cars, with the windows up.

Telephone-type *handsets* are becoming very popular. Available as standard equipment with a growing number of CBs, you can also get one as an add-on necessary from Lafayette, Hy-Gain, and others. They offer you all the advantages of a mike-headset, but don't hang on your head all day. The handsets have switching arrangements that keep the regular speaker on until you pick up the handset, and can also let the regular speaker and handset speaker be on simultaneously, if someone else wants to listen in on the conversation.

SPEAKERS. To keep the overall size down, transceiver manufacturers use very small speakers that not only have limited sound quality and volume, but can be muffled when they aim down at your legs or carpet. A dozen or more manufacturers make extension speakers for mobile use. The best one we've found is the Kar Kriket made by AFS. It uses a patented "working wall" cabinet construction that provides excellent tone from a compact light weight box, and is available in a variety of colors. The Kar Kriket comes with a bracket for mobile mounting (usually under the dash) and is also available with a desk base for indoor use. Most base unit transceivers have good enough speakers, but if you are running a mobile unit off an AC power supply in your house, you could probably benefit from the Kriket. It also comes in handy if you want to listen for a CB call when your transceiver is in another room.

Hy-Gain makes an excellent *amplified* CB speaker that will give your reception a tremendous boost. It connects to the car's battery (or you can wire in an internal battery) and simply plugs into the jack on the back of the CB. Not only does it boost weak signals, but it overcomes noise, and should be considered an absolute necessity for anyone who drives with the top down, or uses a jeep, motorcycle, or snowmobile.

TEST EQUIPMENT. Reception reports from other CBers will tell you if you are getting out, but test equipment will give you an accurate measurement of how well you are doing, and also aid in setting up and maintaining your station.

The most important piece of test equipment you can buy is the SWR meter. Available from 50 manufacturers at prices ranging from $10 to several times that, SWR meters are an absolute necessity for proper antenna tuning. If you can't afford one, borrow a meter from your dealer, or share the cost among some CB friends. They come in a variety of sizes from cigar box to pocket size, and while the absolute accuracy may vary, each model can be calibrated before use, and most will do an acceptable job. Bigger meters are easier to read; smaller meters can easily be kept in your car or tool box.

SWR meters are often combined with other indicators such as watt meters or field strength meters. A watt meter is connected to your antenna line and tells you how much power you are putting out. A field-strength meter is a nonpowered radio receiver that uses a small antenna to pick up your transmitted signals. It doesn't read in watts, but has arbitrary markings that are useful to determine the relative power being radiated by your antenna in different directions. A *modulation meter* indicates your "talk power" and makes it easy to adjust power mikes and to judge how loud and how close you should talk.

You can get a batch of separate meters to make these measurements, but a combination device is handier and less expensive. We recommend the Philmore FSM55 (or the identical Lafayette 99-26387 or Recoton FS9C). These units are rugged, easy to operate, look good sitting on top of a CB, and are compact and rugged enough to keep in your car.

The Shakespeare Defender is a particularly versatile multi-tester, measuring watts, SWR, and modulation, and including a built-in antenna matching circuit and a selector switch to choose either of two antennas.

FREQUENCY METERS are more properly professional test equipment than CB accessories, but a growing number of good buddies are buying them. These units show you in six or more bright digits exactly what frequency you are transmitting on, and can quickly isolate faulty crystals or point to needed adjustments. They are expensive (up to $200) and really only make sense for professional repairmen or CB clubs to purchase, although any affluent CBer will have a great time watching those numbers light up when he hits the mike switch. We've had good luck with models by Palomar and Hufco.

If you really get into servicing (or just love watching flashing lights), you will need an oscilloscope to see the "wave form" of your signal. Leader Instruments makes an excellent unit designed specifically for radio communications equipment. The company also makes a wide range of other test devices, including watt meters and SWR meters, that are the Cadillacs of their class.

PHONE PATCHES. Every driver's dream is to have a mobile telephone, but they don't come cheap, and there is a long waiting period in many areas. You can, however, use your CB to make a phone call from your car, if you speak through a base station equipped with a "phone patch" that connects to the telephone lines. It's perfectly legal (as long as the patch is connected properly, using any "interface" device that the local telephone company requires), and the base station is supervised and not remotely controlled. Radio Shack, Gold Line, and others make budget-priced patches that work adequately, but require *brain-mashing* supervision by the base operator who must monitor the conversation and remember to switch from transmit to receive.

A company called Dorado CB Products makes the most sophisticated and easy-to-use phone patch, which automatically switches the base station into the transmit mode when the person on the telephone says something. It's expensive ($300), but well worth it.

PREAMPLIFIERS. While it is illegal to use an amplifier to boost your outgoing signal, there is nothing to prevent you from using an amplifier to boost the incoming signal. And if everyone uses one, two-way communication is possible over many times the normal range of 4-watt CB. Telco Products and Communications Power make fine pre-amps that attach to the back of your transceiver, and Antenna Specialists makes a unit that goes on the antenna mast. A problem with any of these units, like the old TV boosters, is that it will amplify noise as well as signals, and cannot make one signal stronger without also boosting every other signal on the same channel. A preamplifier, used with a good beam antenna, will give you the most range and intelligibility.

SCANNERS. While not strictly a CB accessory, radio receivers capable of picking up police, fire, and other public safety and private radio communications are very popular among CBers. They'll let you know what Smokey is doing, and also give you a big ear on the world. You can get everything from weather broadcasts to ship-to-shore radios, mobile phones, taxis, planes, and garbage trucks. Frequencies in use fall into three ranges: VHF low (30–50 MHz), VHF high (150–175 MHz) and UHF (450–470 MHz), and you can get radios designed to operate on one, two or all three of these ranges.

People have been listening to these transmissions for years, but the recent boom is tied to the development of the "scanner" radio that automatically switches from one predetermined frequency to the next as soon as the first frequency becomes quiet. This eliminates a lot of dead air, saves the frantic thumb-twiddling necessary when the two parts of a conversation are on two different frequencies, as is common in many safety and utility radio systems.

Scanners are available to work with anywhere from 4 to 24 different frequencies at one time, with the more flexible ones bigger and more costly. Most of the units use a separate crystal ($3–5) for each frequency, and you can find out the frequencies in use in your area from your local precinct station, your electronics dealer, or one of the many law enforcement communications directories available nationally.

You can get a scanner for desk-top use, to hang under your dash, or to fit in your pocket. The most sophisticated unit available is the SBE Opti/Scan. It uses programming cards to select the frequencies you want to monitor and does not need crystals. A good budget pocket-size scanner is the GE Mobile 1 Searcher that does not use crystals, but scans four tunable frequencies. It offers more flexibility than the crystal sets and is less expensive to use, but it might not be selective enough if the frequencies are closely packed in your area. Regency makes a wide range of scanners at many levels of price and sophistication, and Lafayette and Radio Shack have fairly full lines.

The ultimate scanner is the SBE Opti/Scan, a 10-channel crystal-less receiver which digitally derives some 16,000 different radio frequencies. The Opti/Scan takes its name from a unique optical card reader used in programming the device's memory. With a preprogrammed card inserted in the unit, 10 channels can be sequentially scanned for continuous automatic monitoring. Frequencies to be scanned can easily be changed by inserting a different preprogrammed card.

Opti/Scan cards, which are approximately the size of credit cards, can be easily programmed by the user, 10 frequencies per card, to cover any frequencies within the public service, land mobile, marine, or business-industrial FM bands (30–50 MHz, 150–170 MHz, 450–470 MHz, and 490–510 MHz). Frequencies on any of the four bands can be mixed on a single card; and cards can be preprogrammed for use in any geographical area, making it a simple matter to continue monitoring desired channels even when traveling over great distances. Suggested retail price is $400.

The Range Plus preamplifier from Communications Power is an excellent preamplifier. It has a variable control to set just the degree of signal boost necessary (up to 13 dB), and can also be used to reduce your set's sensitivity to eliminate overloading from nearby transmitters. There are pilot lights to indicate transmit and receive modes, it works on both AM and sideband unlike some other preamps, and has a dual power supply for base or mobile installation without external convertors. Suggested retail price is $60.

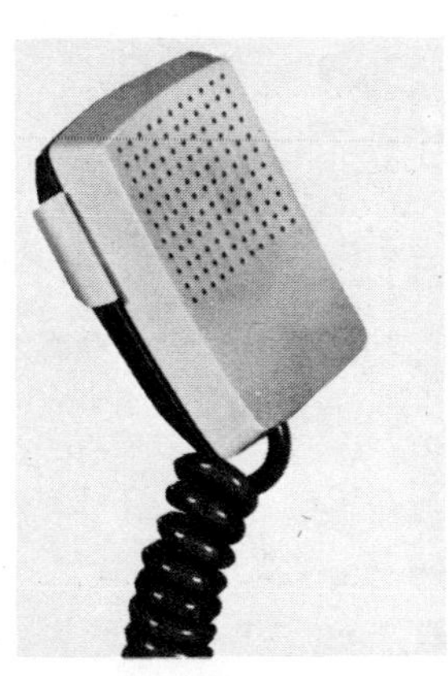

There is no better microphone for normal mobile use than the Turner M + 3. Its "Modu-Gard" internal speech compressor insures maximum "talk power" but prevents overmodulation no matter how loud you speak. And as an added bonus, it keeps your voice at the right level whether you hold the mike at your lips, or much further away. It's also good looking, and well-shaped to the hand. Suggested retail price is $39. A base station unit with similar performance but designed for table-top use is available as the model +3, selling for $51.

The speakers found in most mobile CB transceivers are quite small, and quite limited in sound quality. An extension speaker, such as the AFS Kar Kriket KC35 shown here, will make a big improvement. It not only has better sound, but can play louder than most built-in CB speakers, and can be mounted up high where it won't be muffled by legs and carpeting. It's a good looking unit, available in a variety of colors to go with any car, van, or 18-wheeler. Suggested retail price is $18.

AFS makes a similar unit for base-station use, and a good-sounding, compact outdoor speaker, the KC-45, that is good for mobile PA use. All three speakers work by plugging into the PA, earphone, or external speaker jacks on the back of most CBs.

The "Defender" from Shakespeare is a multipurpose test console that keeps tabs on SWR, modulation, and output, and also serves as an antenna matcher and antenna selector. In short, it

lets you know if your rig is doing anything wrong, and helps you correct it. Suggested retail price is $100.

The Recoton CB 172C is a multipurpose CB tester and performance monitor that combines an SWR meter, watt meter, modulation meter, and field strength meter to analyze just abut any CB difficulty. You can leave it permanently connected and sit it on top of your base transceiver or hang it under your dash; or just take it out and hook it up when you want to make an adjustment. It's accurate, and rugged enough to keep in the trunk of your car. It sells for about $60.

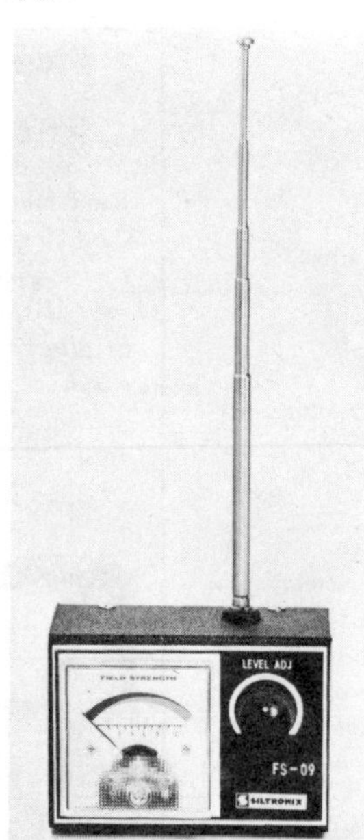

The field strength meter is a small nonpowered radio receiver that indicates transmitter output level. It uses its own antenna, and requires no connection to the CB. It's useful in orienting antennas, tracing cable breaks, and insuring that you are actually getting out on the air. The unit shown here is the Siltronix FS-09. Suggested retail price is $12.

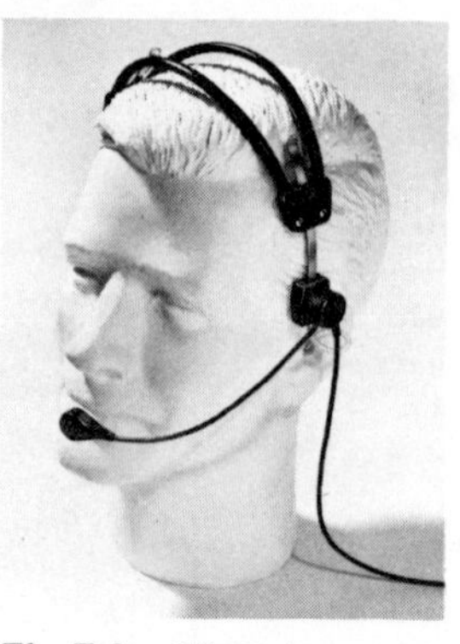

The Telex CB-88 is a sophisticated aeronautical headset adapted for CB use. It is extremely comfortable, and so light in weight that it can be removed from its headband and clipped to the frame of your sunglasses or eyeglasss. The microphone is a noise-cancelling type that is held in the proper position by a pivoting boom. The CB-88 comes with an adjustable mike preamplifier and a remote push-to-talk switch that can be held in your hand or mounted on the car's gear shift lever or any other convenient place. Suggested retail price is $70.

The Leader LB0-310 oscilloscope is one of the most sophisticated test devices available for CB. It performs a multitude of operations, including observing the actual shape of the transmitted radio waves. The price is higher than the usual CB test gear ($270), but the unit makes sense for professional radio technicians, radio clubs, stores, and students.

Pocket-size scanners, such as this GE "Mobile I Searcher," let you keep tabs on the Smokies (or firemen, ambulances, or garbage trucks, for that matter) no matter where you go. The GE unit has unusual tunable station selection that eliminates the need for expensive crystals. It works on 6 "penlight" cells, or with an AC adapter. Suggested retail price is $70.

Noise cancelling microphones eliminate any sounds, like wind, that come from several directions; but pick up sounds, like your voice, that come from only one direction, provided you hold the mike close to your mouth. They were developed for airplane and battlefield

use, but come in very handy in a noisy 18-wheeler or a 4-wheeler with the top down or windows open.

The Telex CB-73 shown here is an advanced design with a built-in battery-operated amplifier with adjustable gain control, used to insure maximum modulation. The noise-canceling function can be defeated when you have the windows rolled up and you want to hold the mike further away. Suggested retail price is $40.

The only piece of test equipment you really need to get on the air properly is an SWR meter, that helps you tune the antenna for maximum output. This Gold Line unit does a fine job, and is small enough to be kept in any tool box. Suggested retail price is $16.

The Shure "Super Punch" 526T is typical of the latest generation of CB accessory microphones, and has some unusual features as well. The internal preamplifier has an adjustable gain control that lets you set up for maximum modulation no matter how soft, or far from the mike you normally speak. A push-to-talk switch can be held "on" with your hand, or locked for long speeches. Another switch lets the unit function with normal switching or automatic VOX units. The mike can be wired for either electronic or relay switching to connect to any transceiver. Frequency response is designed for maximum vocal "punch" and intelligibility on AM and SSB. Suggested retail price is $70.

Mobile antennas can usually be tuned by raising or lowering the whip. Base-station antennas, if they are adjustable at all, necessitate hazardous climbs into the sky. A "matchbox" like this Gold Line GLC/104B, lets you make the adjustments from safe inside your house. Suggested retail price is $10.

Regency, one of the biggest makers of radio scanners, has recently brought out a line of 4-channel models priced at just $90. Three models, ACT-C 4L, 4H, and 4U are available to cover low and high VHF bands as well as UHF. There is a convenient slide-out panel on the top of the case to insert or change crystals for frequency selection, and a built-in speaker and telescoping antenna.

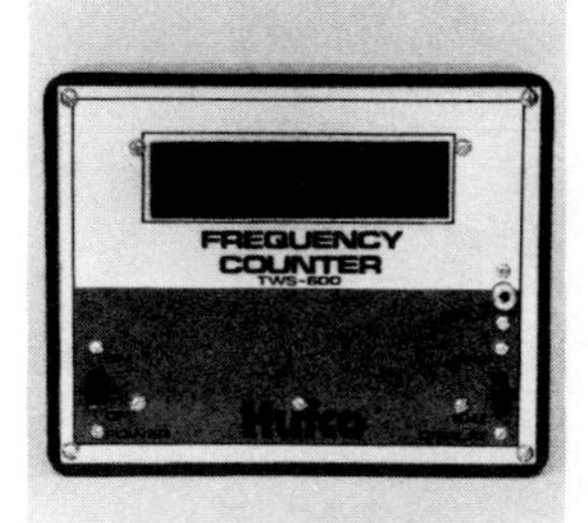

The Hufco TWS-6 frequency counter is a good quality budget frequency counter that will show exactly what frequency you are transmitting on. It works on either 12 volts DC or house current, and is small and tough enough to be kept in a tool box, although most people will probably want to leave it set up at all times so they can watch the flashing numbers. It's available as an easy-to-build kit for $70, or fully assembled at $100.

The new Superex CB-1025VX is a mobile headset with microphone that comes with a voice-activated transmitter switch (VOX), so you will never have to take your hands off the wheel. Suggested retail price is $100.

6.

How to Select and Service Your Own Transceiver

CB Shopping Suggestions

Until recently, consumers in the market for CB equipment had to choose from a limited number of retailers. You could find a CB set tucked away in a dusty corner of an electronics parts distributor, where the guy behind the counter was more contemptuous than knowledgeable; or you could buy from one of the national retail/mail order electronic giants: Olson Radio, Allied Radio, Radio Shack, or Lafayette Radio.

The salespeople at Lafayette and the Shack were knowledgeable and usually patient (except on a busy Saturday afternoon), but your choice of equipment was limited to the store's own "house brand," and because these places were oriented to the do-it-yourself electronics hobbyist, installations were usually unavailable.

Most locations also had a hobbyist-turned-dealer selling from his basement or garage. This pro knew what he was talking about and had plenty of time to help, explain, advise, and comfort, but he carried very little inventory, had no charge accounts, and when repairs were necessary, would usually be unavailable.

Today the situation is radically improved. Industry forecasters predict sales of nearly ten million CB units in 1976. This means that knowledgeable businessmen are now selling CB, offering excellent service, broad selections, and fair prices.

Mail Order

Postmen deliver a lot of CB gear. Lafayette and Radio Shack do a large amount of the total mail order business, but with their ever-increasing number of stores and franchises, they are relying less and less on direct marketing. Their catalogs, however, offer a wealth of information, and are well worth studying to get an idea of the kind of equipment available at different price levels, whether you end up buying from them or not.

Other electronic specialists, such as Olson, Electronic Distributors, Inc., and some mail-order hi-fi companies, sell many items, but they don't carry as broad a line as the Shack or Lafayette and the catalog entries are less detailed. However, their prices, particularly on discontinued models, can be very good.

Traditional mail-order firms, such as Sears, JCPenney, J.C. Whitney, credit card companies, and catalog showrooms, are now into CB with as much enthusiasm as they put behind other hot items. Company attitudes, policies, and services vary greatly. Sears sells Johnson CB gear at full price. Consumers Distributors sells Johnson at big discounts. Penney has its own brand at competitive prices. The credit card companies advertise unknown brands. J.C. Whitney's catalog shows what look like national brand items, but they don't identify the brand names. Prices are on the low side.

Should you buy by mail? Well, it depends. Obviously it takes time; you have to wait for your order to arrive, you have to wait for repairs, and you'll have to work out your own installation. But

navaho
• ALL CRYSTALS • ALL CHANNELS

if there's no dealer in your area, or if the locals don't have the brand you want or charge too much, it might be the best approach.

By the way, don't forget that your local Sears or Penney's probably stocks some CB equipment. It is usually in the TV/radio section or auto accessory area.

If the store has an auto-repair department, you can probably get the unit installed on the spot. Installation prices are usually cheaper than at an electronic store. But inquire first about the equipment on hand and the experience of the installer. (For a starter ask: "Do you have an SWR meter). As a rule, if there is a mechanic on hand and who's into CB, you'll get a good job; if there's no such resident specialist, the CB will be slapped into your car by the guy who does AM/FM radios and window defoggers. He's probably not very knowledgeable about CB, doesn't have the special tools, and possibly, he'll set you up so you're only putting out one-third the normal power you would be generating if the job were done more professionally.

Department Stores

Most consumers see their first CB set in a big department store, simply because they are more likely to be in such a place than in one of the electronics specialty stores. The problem with department stores is that the people who sell the CB equipment spend most of their time selling color TV, washing machines, cameras, or calculators, and aren't particularly knowledgeable about CB. Be very careful! And be aware that since most department store salespeople work on commission and would much rather sell you something that they will profit from, instead of simply catering to your needs. Be sure to read up on the literature before you actually purchase your rig.

Selections vary, but typically a department store will take in one line of three to five models to test the market, and if the goods move off the shelves reasonably fast, the store will bring in a few more brands. Accessory selections are severely limited, and prices are high. Transceivers are seldom discounted, but this will probably change in the near future when supplies increase and the stores include CB in their big sales.

When it comes to installations, you're strictly on your own. Even department stores with auto sections seldom install CBs, and the most cooperation you can expect is a referral to an installer.

Why would anyone buy CB from a department store? Well, you can use your charge account. Return privileges are liberal. And you might find something on sale. However, remember that department stores generally aren't the place to go for professional advice.

Appliance Stores

These wheeler-dealers have gotten into CB simply because it is

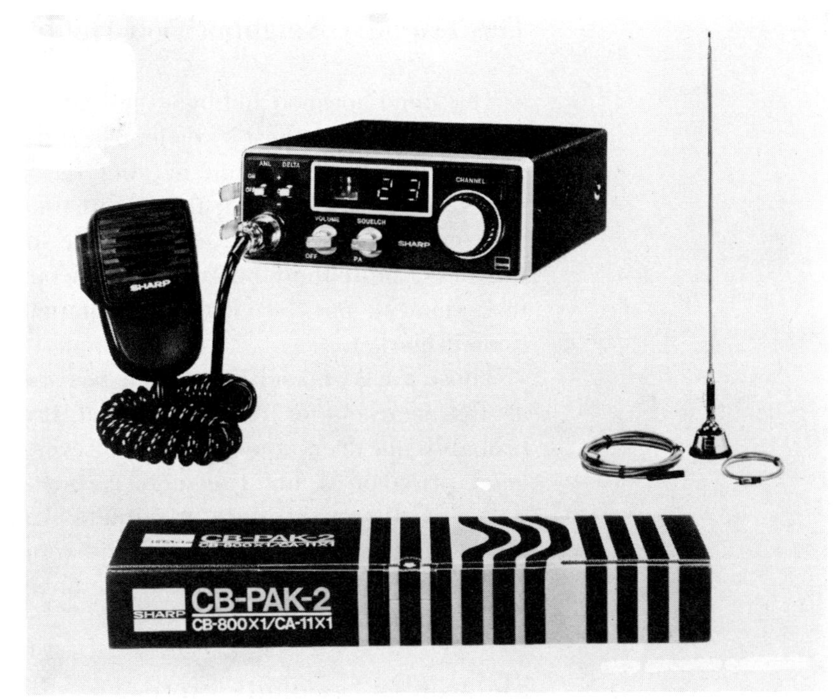

The Sharp CB-Pak 2 makes shopping a cinch. You get a good quality transceiver (with electronic digital channel readout, ANL, delta tune, and PA) plus an antenna, hardware, and cable. And everything comes in one box.

the latest "hot" electronic product. Some of the salesmen have taken the time to digest the manufacturers' literature, and diligently study the trade papers, but like in the department stores, there are many who really are not knowledgeable at all. Again, it pays to know something before you enter one of these stores, as you just might find a very good buy there.

Selections in the appliance stores are limited, accessories seldom stocked at all, and the price seems to fluctuate greatly. However, it is possible to get a good buy when business is slow and the salesman is eager to move goods at any price (most any weekday morning); and there are frequent sales, but the promoted items are often brands nobody has ever heard of. If you see a special price on the Irvingsonic X4, or Whoopietronic 23, run away.

Don't expect an appliance dealer to install a CB in your car. But if you're buying a base unit, you just might find that the appliance dealer (who is accustomed to putting up TV roof antennas) can take good care of you. Make sure the installer has an "SWR meter" and knows how to use it.

Auto Accessory Stores

Those glass-walled storehouses of Super Sport ashtrays, Rallye mirrors, and Monte Carlo mufflers will be happy to sell you a CB set. You'll find them displayed near the tape decks or fog lights, and you can usually get an installation on the spot. Technical knowledge varies greatly. The bigger national and regional chains have been holding CB training seminars recently for their employees, and most local stores have at least one CBer on the staff who knows what's what. But a lot of the accessory houses do nothing more than take your money and hand you a box.

The Friendly Neighborhood Hobbyist

The neighborhood hobbyist selling from his house has largely disappeared from the CB scene. He did his part to propagate the radio faith in years gone by, but has now lost out to the real stores. In a way it's sad; these part-timers did a lot for CB users and makers (Browning, one of the oldest companies in the business, built their business by encouraging CBers to sell from their homes), but they just can't compete in today's more sophisticated market.

Those guys who sell from their basements today (or from tables at flea markets or from trucks at the side of the road) are probably not CBers and could be thieves or fences. You can get a good price on a hot transceiver, but even if your conscience doesn't bother you, there are practical disadvantages. The units have serial numbers, and cops and repairmen have been known to check the numbers against "hot sheets." It pays, in the long run, to be honest.

Electronics Specialty Stores

The neighborhood radio parts dealer of 20 years ago has either gone out of business or evolved into a broad-lined electronics specialist, often carrying everything from a ten-cent resistor to a thousand-dollar amplifier, as well as car stereos, TVs and CB. This is the place to get good advice, plenty of information, competent service, proper installations, huge selections, and good prices. In some areas you can find dealers who sell nothing but automotive electronics. This may be a dealer that started with car stereos and expanded into CB.

National Chains

The regional chains of electronics discount stores, such as Tokyo Shapiro and Arrow Electronics sell CB and usually have good selections of transceivers and accessories, most at promotional prices. Few of these stores, however, do installations, and the salespeople seem to be more involved with hi-fi than CB.

The two principal national chains, Lafayette Radio and Radio Shack, sell only their own brands of CB gear. But don't be afraid. These private label items are a far cry from the old "off brands." Radio Shack's "Realistic" label, for example, should probably not even be considered a private brand, because it's available in more than 4,000 stores.

Lafayette and the 'Shack have been selling CB since the very beginning, have had a good number of product firsts, and are generally well regarded for technical sophistication and durability. One of the authors of this book started with a Lafayette HE-20, and that unit is still going strong after 15 years. He also has Realistic gear nearly as old that still works just fine. Today both companies have broad lines of CB gear, and offer fair prices

CB stores take many forms. This true "mobile unit" can be found along the highways of upstate New York. He carries a wide variety of merchandise and installations are available on the spot.

(with frequent sales), liberal return policies, and quick repair service in most communities.

Buying and Selling Used Equipment

You can often get a good buy on used equipment bought from fellow CBers at Coffee Breaks, Jamborees, or Swap Meets, but be careful. Try to buy from someone you know or from someone who will accept a check, and be sure to try out the unit before you buy it. If it works then, it will probably work later, but let it run for 10 minutes or more so it can heat up.

Some dealers will accept your used equipment as a trade-in on new gear, but you will be lucky to get even half its original price. If you sell privately, you can expect 60–70% of original price if it's only a few years old and in good shape.

CB Security

While some of us are still fortunate enough to live and work in communities where the locked door is an uncommon aberration, most Americans are not so fortunate. Crime is increasing everywhere, and CB is a prime target, because it is light in weight, easily concealed, and in great demand. The antenna on top of your car or house is a neon sign attracting thieves, and while we can't expect you to transmit without an antenna, there are protective steps you can take. If you travel in high-crime areas, you can get an antenna such as the Antenna Specialists MR-264 or Realistic 21-930 that looks like a car-radio antenna and works for AM and FM as well as CB. It won't work quite as well as a real CB antenna, but it might save your set.

When you park your car, remove the antenna unless you will be able to keep watching it. Most units can be unscrewed in a

A CB specialist like the M & K Mobile Center in Flushing, New York, stocks a dozen brands of transceivers and every imaginable antenna and accessory. The salespeople are knowledgeable, and have tried out everything they sell. Installations are available, problems are solved quickly, and prices are always fair and often a bargain. If you can find a store like M & K in your community, you will be in great shape.

few seconds and stowed in the trunk, and you can make the job even easier by using a quick disconnect mount, such as the Valor 103 or Antenna Specialists M54. (Don't forget to retune your antenna with an SWR meter to compensate for the extra length added by the quick disconnect unit.) Even with the antenna removed, the antenna base will still be a tell-tale sign to any thief who is studying carefully. A magnetic mount antenna, such as the Turner SK900 can be completely removed when you park, leaving no hardware exposed, yet will stay upright at high speeds. If it is possible to still use it safely, mount the transceiver where it can't be seen from the outside, maybe in the glove compartment, center console, or far up under the dash. Remote control units, such as the new Royce 1-606 and Radio Shack One Hander let you keep the bulky transceiver in the trunk or some other out of the way place, and use just a tiny, easily concealed "control head" on the dashboard.

Many people camouflage their CB installations. You can get a unit that is the same color as the dashboard so it blends in (black, silver, and wood-tone units are available, or you can cover up the CB with dark cloth or an empty Kleenex box "shell" when you park. Remove the microphone if you leave your car in an unattended parking lot. It's certainly a good idea to remove the CB whenever possible, and there are a lot of slide-out mounts, such as Lafayette's number 42N38002, which make it easy to yank out the CB to carry with you or lock in the trunk. Some CBs, such as the new Craig line, have built-in slide mounts, and look quite nice.

If you insist on leaving the CB in the car, by all means beef up the flimsy bracket that came with the set. Shur-lok makes a number of super-strong mounting systems for use under the dash or on the transmission hump, and SST makes locking brackets.

As the car stereo makers have shown, it's a lot harder to steal something that's inside the dash than under it, and you can now get in-dash CBs from JIL, Panasonic, and Kraco that fit in the hole where the original car radio was. These are combination rigs, including radio and/or tape deck as well as CB, and they do a surprisingly good job in all modes. Their only real disadvantage is that when one function doesn't work, the whole unit goes out for repairs and you're left with nothing to listen to.

There are hundreds of burglar alarms on the market. Some of them are specifically designed for CB use, but you may as well get a complete system that protects the whole car. If you need further protection, one company markets a tear gas gun that releases its fumes when the CB is removed from the dash.

At home, use common sense. If you live in an apartment house, try to hide the cable that goes from the antenna to your apartment. Get a good lock, a good alarm, and don't talk to strangers. Keep a record of your set's serial number and make sure your insurance covers CB, (in some states there is an extra charge).

One of the greatest disappearing acts we've seen is performed by the new Shur-Lok "Flip-Flop" antenna mount. When you're driving, your antenna is held straight up. When you park, you raise the trunk lid and flip the antenna inside, safe from would-be thieves. It can be used with almost any antenna, and installs with two small holes in the trunk groove. Suggested retail price is $12.

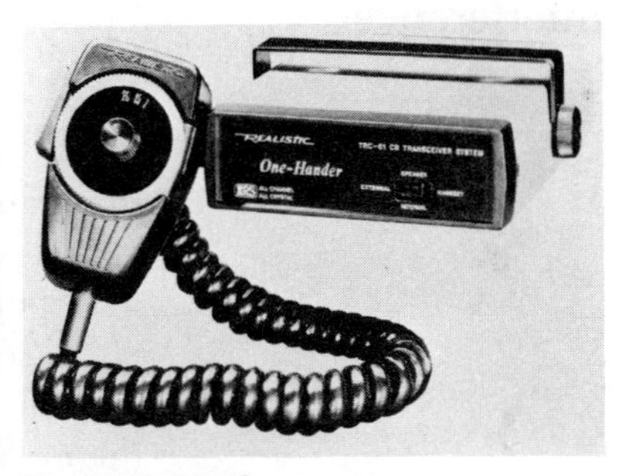

The Radio Shack "One-Hander" is a strange radio. All the operating controls—channel selector, volume control, squelch, and even a small speaker, are built into the microphone. That way it's particularly safe to operate, since you don't have to take your eyes very far off the road, and it's almost completely safe from thieves. You can mount the main transceiver section in the glove compartment or under the seat, and tuck the microphone away when you park. (For optimum security, use it with a removable or camouflaged antenna.) Performance is comparable to conventional CBs. A switch on the transceiver selects either the speaker in the microphone, a larger one in the transceiver, or an external one. $150.

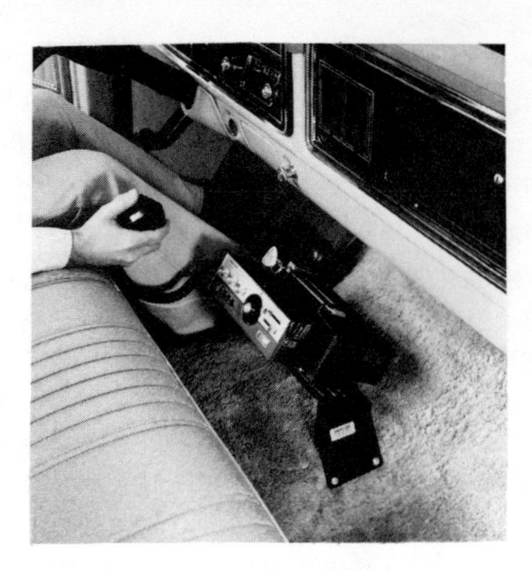

Shur-Lok makes a wide range of extremely rugged and well thought-out security brackets and mounts for CB. The hump mount shown here is great if there is no room under the dash, or if you want the CB mounted between a driver and passenger who both use the transceiver. It attaches to the transmission tunnel with special one-way, high-strength screws that look like they would take an explosive to remove. The CB is secured with a high-strength pick-resistant lock. Suggested retail price is $30, which is considerably cheaper than a new CB.

Eliminating Interference

CB radios receive a lot more than voices. The airwaves are loaded with snaps, crackles, pops, and whistles, only some of which you can eliminate. In mobile CB, the chief cause of interference is the car engine, both your own, and every other one on the road. Most CB sets have ANL circuits or noise blankers that can take care of the trouble, but sometimes you need a more complex cure.

A popping sound that increases in speed when you accelerate is caused by the ignition system and can usually be eliminated by switching to resistor spark plug cables if your car does not have them. Most cars do, but certain high performance models have "solid" spark plug cables. These can be replaced with MSW cables available from Time Machines, Inc., that provide good suppression with no detectable change in performance, or you can connect spark plug and distributor suppressors sold in kits like the Radio Shack #21-502. A whining sound that increases in pitch as you accelerate is caused by the generator or alternator, and can usually be stopped with an alternator/generator filter like the Gold Line 1080 or Cornell-Dubilier CB100, or a filter attached to the voltage regulator, (like the Radio Shack #21-507).

A CB base station can pick up pops every time a switch is turned or a thermostat clicks, and while these noises are not much of a problem most of the time, they can interfere with long-distance communication. The Cornell-Dubilier CBBS-1 is very effective for such noises, but it costs $20. If there is only one popper in your house, you might be able to isolate it with Cornell-Dubilier's $4 plug-in filter, CBBS-2. If neighbors hear your voice on their TV or see strange lines when you transmit, the fault can be yours or theirs. If you determine that only some TVs are affected, they should be equipped with a high pass filter, like the Gold Line unit.

If you interfere with every set, put a low pass filter, such as the Telco Channel Guard, in your CB antenna line.

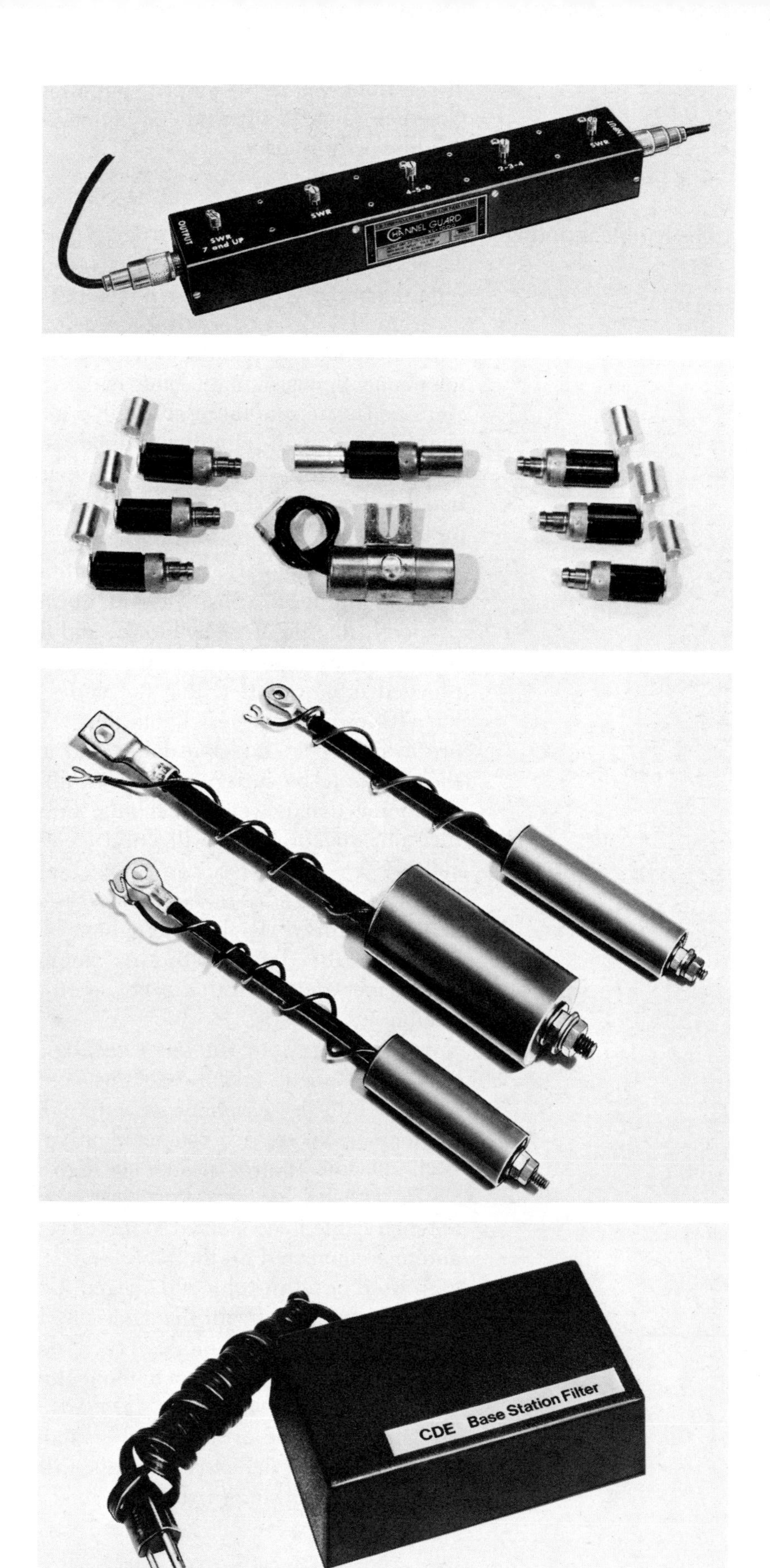

The best TVI suppressor we've tried is also the best antenna matcher. The Telco Channel Guard low pass filter reduces radiation above the Citizens Band by 100 dB, and is certain to restore domestic tranquility and make peace with your neighbors. It has five separate—but easy to use—controls that pinpoint and knock out interference on any T-V channel, and it provides as perfect an antenna match as we've ever seen, with the absolute lowest SWR for minimum wasted power. Suggested retail price is $35.

Noise suppression kits, such as the 21-502 from Radio Shack, eliminate the noise generated by the ignition system in your car. Resistors plug onto the top of each spark plug and into the center of the distributor, and a capacitor is provided to knock out other noise. Retail price is $4.50.

Cornell-Dubilier makes a variety of filters to attach to the generator or alternator of any car to eliminate whining interference noise. Prices range from $8.25 to $18.90, depending on size.

The Cornell-Dubilier Base Station Filter goes a long way toward eliminating all those pops and clicks that happen whenever anyone in your home flicks a switch or the refrigerator or fish tank heater goes on. Suggested retail price is $20.

If you think *you're* the culprit, put a low pass filter like the Teiko Channel Guard in your CB antenna line and most of your troubles will be over.

CB Troubleshooting

Most CB transceivers are rugged and reliable—there a lot of 15-year olds still used every day — but when they conk out, repairs can be very expensive (seldom under $25) and you might be without your set for a long time. Although the law restricts most CB repairs to people with FCC First or Second Class Radiotelephone licenses, and some makers' warranties are voided if you open your set, there are still a lot of checks you can make yourself that will eliminate unnecessary bills and delays by isolating any *external* troubles. Compare your symptoms to those listed below, and then try our suggested cures to pinpoint the problem.

SET DOESN'T TURN ON. If it's a mobile unit, are you sure your hot and ground leads (positive and negative wires) are connected properly, that no fuse has blown, and the power cord hasn't come out of the back of the set? If it's a base unit, are you sure it's plugged into an outlet, that the outlet is live, and the set's fuse hasn't blown? Little test lights that work on home or car voltage are available for under a dollar and are very helpful in making these tests to be sure power is available to supply the CB. If you determine that power is reaching the set and the set still won't turn on, you probably will only need a new power switch, that sells for a few dollars.

YOU RECEIVE, BUT YOU CAN'T TRANSMIT. Is the microphone plugged in securely? Is the microphone volume control (if any) set high enough? Try substituting another mike from a similar transceiver and see if that gets you on the air. If it does, you just need a mike repair.

YOU TRANSMIT, BUT YOU DON'T RECEIVE. Are you sure the squelch control is "open" and the volume control up? Is there anything plugged into the earphone or external speaker jack, cutting off the main speaker? Are you accidently switched into the PA mode? Is the RF gain control (if any) set high enough?

YOU RECEIVE AND TRANSMIT, BUT NOT WELL. Are you sure the antenna cable is connected to the CB set? Are you sure there is an antenna connected on the other end? Is the SWR too high? Has a high wind or falling object damaged the antenna? Try substituting a different antenna with the same CB. Try substituting a different CB with the same antenna. Try a different cable. Check the connectors for shorts or open connections.

YOU TRANSMIT AND RECEIVE, BUT NOT ON ALL FREQUENCIES. You probably have one or more bad crystals. Your dealer can probably replace any defective crystals without charging you for his time if you give him a list of the good and bad channels.

The ultimate CB test instrument may well be the smallest. The "Port-a-Test," made by Telco Products Corp., is half the size of a band-aid box, and performs a huge number of functions.

Inside the unbreakable nylon box is a signal generator that puts out both audio and radio frequency signals, plus a light-emitting diode test light circuit.

You can feed the signal into your transceiver to see if it is receiving properly and to verify that accessories like preamps are working. You can use it to calibrate one S-meter against another. You can test speakers and microphones. You can check for broken microphone, speaker, and antenna cables. You can check switches and plugs. You can verify polarity on power supply connections. You can check to see if batteries or AC outlets are live, you can use it to trace wires and cables that are buried under the ground or behind walls. And a hundred other things.

The signal generator circuit can be switched to provide either a steady 1000 Hz tone, or a tone that shifts from 500 to 1000 Hz every half second, and sounds a lot like the sirens on European police cars. You can feed it into an amplifier at home for an emergency alarm, or use it with a car PA speaker. Suggested retail price is $30 with a leather holster.

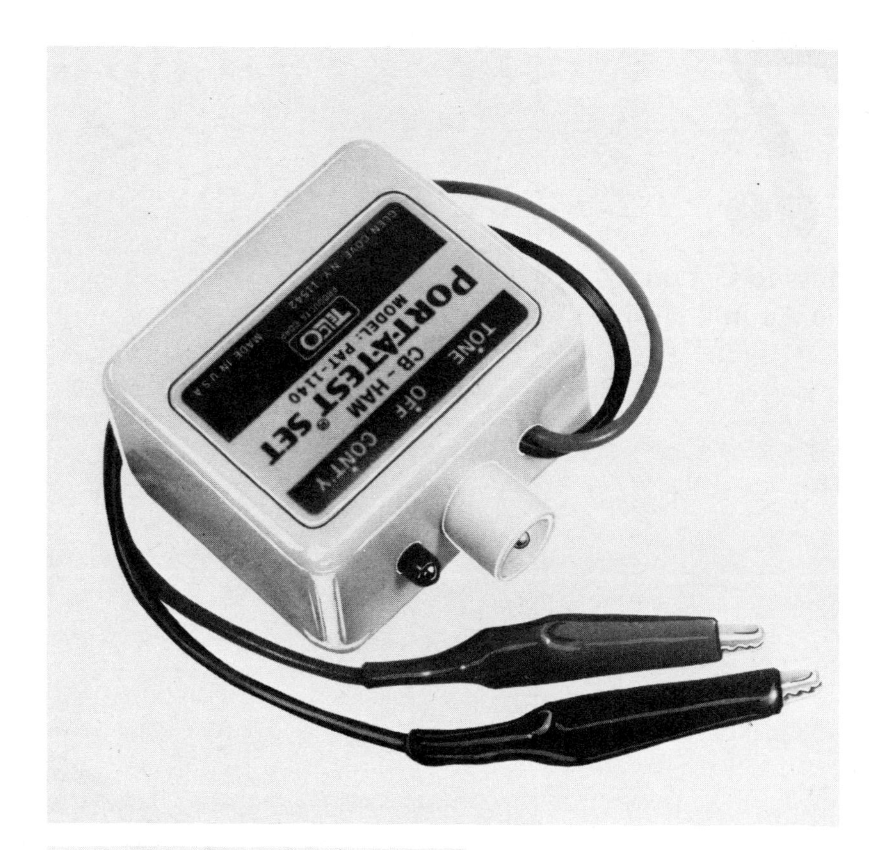

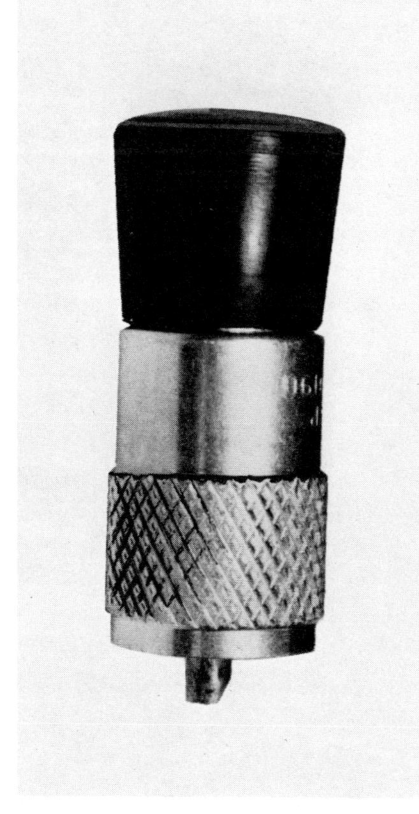

If you are making adjustments or measurements on your CB that necessitate keeping the microphone keyed for a long time, it's a good idea to use a "dummy load," such as the one pictured, in place of the antenna. That way you won't tie up the channels. Dummy loads sell for $3–5. Some of them have integral lights that flash to show relative output.

7.

How to Get on the Air in a Hurry

686
Atla
WHITE

Your License and the Law

You can't legally get on the air in a hurry unless you have an FCC license and call letters, or you're a member of a club or organization whose FCC call letters you have permission to use. But nobody, not even the FCC can stop you from listening in.

Getting a license is easy. Fill out an application (you will find one included in the Appendix)and mail it to the FCC along with $4. A computerized processing system went into operation in late 1976 and your license should be returned about ten days after you mail the application. You can put that time to good use by listening on the air and spending some time with our CB Dictionary, looking up words and phrases you hear on the air.

After you receive your license, the FCC requires that you keep it near your transceiver; either in your mobile unit or at your base station. It's illegal to operate without giving out your call letters at the beginning and end of each transmission. If you fail to do so, the man from the Friendly Candy Company will track you down with direction finding equipment and hand you one of his famous candy-pink and white striped summonses. You are also required to keep a copy of any code you may want to use near your transceiver.

Your Equipment

A CB transceiver is not much more complicated to operate than a regular AM radio and setting it up is a simple task. First, read the manufacturer's instructions and then connect your rig to a power source and your antenna to your rig. *Be careful!* An improperly connected antenna can burn out a transceiver when you are transmitting. Make sure the antenna you are using was designed to be used with CB equipment.

Once your equipment is properly connected, turn it on and set the volume control to a middle position. Then adjust the squelch control until you hear something. Rotate the channel selector until you pick up a conversation.

Modulating

Now you're a rubberbander, ready to modulate. If you haven't already done so, spend some time listening to different channels. You may notice that some channels in your area have distinct personalities or that certain channels have been "taken over" by an ethnic or social group. In all states, Channel 9 is reserved for emergency communications, while Channel 11 is the official calling station used to establish contact and then abandoned when you switch to an agreed-upon channel for modulating with your partner. In most parts of the country, Channel 19 is the truckers' channel and Channel 13 is the unofficial marine channel.

When you're ready to modulate, wait until the channel is quiet or there's a pause in the conversation and then say, "Break (channel digits)," or "Breaker by, on (channel digits)." For example, if you are on Channel 15 you would say, "Break, 1-5," or "Breaker by on 1-5."

If you're lucky, shortly after you've broken in someone will respond with: "Come on, breaker," or "Go ahead, breaker." Now

The CB'ers Creed

As a Citizens Band Radio Operator, I recognize my obligations:

To the Federal Communications Commission and The United States of America, who believe that I am sufficiently mature to be entrusted with the ownership, control, and operation of a radio transmitter.

To my neighbors, who trust their lives and safety to my skill and judgment during times of emergency.

To my fellow Citizens Band operators, who depend upon me to follow established good practices, procedures and courtesies.

To discharge these responsibilities, I will at all times observe the highest standards as a Citizens Band operator.

I will never knowingly cause interruption to Citizens Band stations engaged in communications.

I will be careful to avoid generating interference to radio and TV receiving equipment, and will endeavor to locate and eliminate interference to any such equipment which may emanate from my station.

I will transmit only to pass necessary and substantive messages.

I will make all efforts to make full and proper use of Channel 9, the National Calling and Emergency Channel.

I will aggressively maintain my proficiency as a Citizens Band operator and keep abreast of electronics and communication developments so that my operation, which largely depends on such knowledge, may be of the highest order.

I will conduct myself on the air to reflect credit upon myself, the Citizens Radio Service and my country.

I will constantly strive to keep my standards high.

I pledge adherence to these principles so that I may contribute my part to more efficient radio communications, and advance the dignity of the Citizens Radio Service.

____

Operator's Name

identify yourself with your call letters and handle. Here's where some rubberbanders have trouble or suffer "mike fright." What should they talk about their first time on the air? The truth is, as you may have already discovered by listening, you can talk about anything from baseball scores to your favorite brand of beer. Ordinary English will probably get a response, although some CBers, and truckers in particular, are reluctant to talk with rubberbanders. Most CBers develop a country trucking drawl when they modulate, but this shouldn't be forced. It will come naturally in time. The most important thing is to be honest. If you're sincere, CBers should respond.

What If Nobody Answers?

It's rare, but not unheard of, that a rubberbander will have a difficult time getting talked to. Sometimes CB cliques feel they "own" a channel and refuse to talk to outsiders. In cases like these, you may not even be able to determine if you are putting out a signal at all.

If you are having a difficult time striking up a conversation, get to know some of your CB neighbors on an "eyeball" basis. Attend a CB club meeting, coffee break or jamboree in your area. As you get to know some CBers, make arrangements to meet them on the channel of their choice at a time that is convenient for everybody.

If this doesn't work, or if there isn't a club in your area, go to a CB store and wait for another neophyte CBer to come in and look at equipment. As an "experienced" CBer, help him out. Suggest transceivers and antennas. Tell him about your model. Then do him a real favor. Arrange for him to call you on your favorite channel at a time that is convenient for you. In this way, you can begin to generate your own network of CB friends.

Despite initial difficulties some CBers encounter, it shouldn't be long before modulating on the CB becomes as natural as talking on the telephone.

OPENING GAMBITS GUARANTEED TO GET A RESPONSE

"Did everyone see that Smokey in the bushes? I think he's taking pictures, come-awn."

"Hey, cottonpickers. I gotta YL hitchhiker here looking for her next ride. Anybody heading her way, come-awn?

"Anyone got a copy on this here lonesome lady?" (This will only work for a female.)

"I'm about to put my boots on–if somebody's too close, give a shout."

In our society, most of us are stuck with our given names or with nicknames we may actually dislike, but in many cultures when people come of age, they choose new names for themselves, names they like, identify with or that challenge them. Choosing a handle, the name CBers use both on the air and when they get together at social events, is participating in a true initiation rite.

Creating a handle for yourself provides you with an opportunity to chose a name you like, formalize it and stake out an identity for yourself in CB land. If you're like many CBers, you'll discover that being called by your handle can have an uplifting, almost liberating effect and you'll probably prefer using your handle most of the time.

Getting Hold of Your Handle

For the most part, CBers choose handles that extend real or imagined aspects of themselves. Often they take the opportunity to make light of a unique character, personality, or physical trait. For example, an extremely tall CBer in California calls himself Bean Pole while his friend, a slightly overweight, jolly fellow with a deep laugh and thick glasses, is Toad Frog. Frito has been eating potato chips everyday for the last ten years of his life and Ribbon Wrangler is a truck-driving poet. Easy Money is famous for his clever projects and schemes. A transplanted midwesterner now living in California calls himself Proud Oakie.

Some CBers take themselves and their handles even more seriously. Macho males go for names like Road Ranger, Channel Hog, Track Master, Leatherneck and Top Line. Female CBers, beavers, take handles that range from the suggestive — Red Streak, Mama Cat, Lacy Lady — to the more lighthearted and sublime: Paper Doll, Lost Love, Bambi, Bubbles and Little Deer. At any gathering of CBers you'll find names that are pure fun—like Fud Pucker, Golly Womper, Paddlefoot, Coat Hanger, and Oakie Odd Ball.

It might seem as if there were an unlimited number of handles to choose from. Unfortunately, that's not the case. CBers treat their handles as possessions and, consequently, can get extremely upset if someone else turns up with the same name. A thriving business of handle registries has sprung up to provide solace, if not solutions to these social crises. For a nominal fee you can register your handle and "protect" it from infringement in a given geographic area. Each year these registries publish handle directories, establishing a semblance of proprietary right to those whose handles are listed.

But the choice of a handle, even if limited, is yours and yours alone. Be whoever you dare and choose wisely.

UNITED CB REGISTRY

This is to certify that the CB Handle of

Blue Goose

has been registered with UNITED CB REGISTRY and will be entered into the records permanently as assigned

to Donald G. Burton of Houston, Texas

Registration of assignment above is effective as of February 6 1976

Signed E.L. Rosa Secretary

Signed J.W. McAllister President

UCBR

REGISTRATION NUMBER A 12823

8.

CB in the City: Urbanites Make the Airwaves Their Own

FINA
AMERICA
CLUB
ORT WORTH, TEXAS

The Urban Airways Abound in Unusual Approaches to the "Truckers Toy and Joy"

The "Channel Master" sat hunkered down in the front seat of his old Plymouth. He'd been there for over two hours and the heater, working overtime, was only meeting the chill of this early February morning half way. Wiping off a patch of steamed-up window, he glanced over at the old clock above the entrance to the darkened school building across the street and quickly stirred himself into action. The CB was on, volume up, but only static crackled out of the tiny speaker as he pushed the talk button on his mike. Even though he routinely asked for a 10-13, "Channel Master" wasn't looking for any ordinary traffic report. As he saw the big orange bus rumble down the street toward him and pull to a stop in front of the old school, the "Channel Master" knew that most of what happened next would be headline news across the country. He wasn't wrong.

"Yesterday's school riots in South Boston," the Associated Press quoted Boston Police Chief Robert DiGrazia, "were engineered by unlawful, antibusing fanatics who used citizens band radio to coordinate an organized assault on police officers and public school officials." DiGrazia and Boston Mayor Kevin White called for an immediate investigation by both the FBI and the FCC.

CB is as popular in the cities of America as it is out in the hinterlands and on the open road. But because of the prevailing conditions, its uses in an urban environment are sometimes startlingly different.

There's little point, for example, in really asking for a 10-13 (traffic report) when everyone knows the street today is jammed bumper-to-bumper for miles in all directions, and no one worries about losing green stamps to a picture-taking Smokey when the average speed is just a single nickel. Moreover, the 4-ft mobile antenna that carries a country-mile out on the interstate may have trouble making it down the block and around the corner in a city filled with 50-story buildings.

CB is very effective, however, for security and for socializing. In New York, Atlanta, Chicago, and Philadelphia, for example, there are apartment house doormen, auxiliary police, hired guards, block association patrols and even little old ladies who sit by their windows all day, all equipped with walkie talkies that connect them to neighborhood REACT bases (which are often located right inside the precinct stations or local firehouses). In Boston, New York, and Washington, D.C., taxi drivers carry CB units in addition to their regular radios so they can alert emergency teams to trouble in the streets or join in searches for stolen vehicles or criminals on the run. Whenever a particularly shocking crime—a kidnapping, for example—occurs in Dallas, CB-equipped vigilante teams will form in minutes, mounting a strong citizens' attempt at enforcing order and tracking down the perpetrators. Sometimes, these CB vigilantes develop valuable leads for the professional law enforcers. Once in a while they actually come up with the culprit.

Herman Weinberg had little use for the law in the Bedford-Stuyvesant section of Brooklyn where he lives. Two years ago his

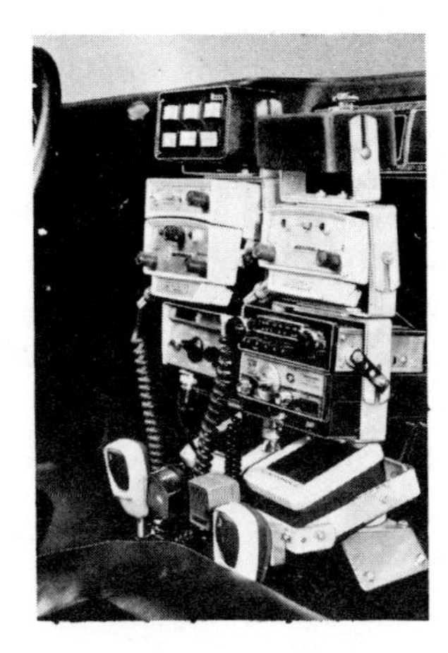

teenage granddaughter was mugged and raped in an apartment elevator on her way to visit Herman's wife, Heide, who has been bedridden for almost a dozen years. Two men shut down the elevator, attacked his granddaughter, and then escaped through the roof of the building—all in broad daylight and within sight of a half dozen eyewitnesses, including two white police patrolmen assigned to this predominantly black precinct. Immediately afterward, Herman joined the Vigilantes, a semi-official organization of local Bed-Stuy residents who couldn't afford the luxury of private guards in their buildings or on their streets, but who nevertheless decided to do something about the situation. Twelve hours a week, for two six-hour shifts, Herman Weinberg leaves his dry-cleaning business, dons a white warm-up jacket and hardhat, takes his Midland portable CB unit and a 12-inch tire-billy and, with another member of the Vigilante organization, patrols a six square block area on foot. Herman can't undo the crimes perpetrated on his granddaughter, but he and the other Vigilantes have insured that their neighborhood's crime rate has been cut in half.

Precinct figures show that anti-bussing boycotts in Louisville last year were all engineered by CB, and school strikers in Pittsburgh and Chicago (who couldn't wait for the "all-news" AM radio stations to tell them what was happening) depended upon their CBs to stay informed.

In many cities, particularly where it is no longer considered safe to be on the streets after dark, CB has become the modern day malt shop, replacing the candy store and pool hall as the neighborhood hangout. Instead of "Meet me downtown at 6," teenagers and adults are now saying "Meet me on 8 at 6."

Today's kids plead with their parents to let them stay *on* late, instead of *out* late. Just as each city neighborhood once had its favorite gathering place, each now has its favorite channel, and anyone not from the neighborhood is either considered an outsider or a troublemaker. Generally your break into a busy neighborhood frequency will bring you nothing but dead air. Pity the poor guy from the wrong part of town who unwittingly invades someone else's CB turf, transmitting on a channel that "belongs" to another neighborhood. When some kids from New Jersey, for example, began using the channel claimed by the Bronx recently, more than a score of roof top antennas were broken in the resulting battle.

We even have ghettos on the airwaves. In New York, the only language you'll hear on Channel 22 is Spanish. In Washington, D.C., black people favor Channel 3 or 4. Channel 8 in Boston is pure Portuguese. There are special purpose channels, too. On the highways outside San Francisco, Channel 19 will get you traffic reports; but when you get into town and want to set yourself up with a date, try the Singles Bar frequency. Brad Whitmore from the Carmel Valley did and couldn't believe what happened. He rolled into Friskie City one Friday evening, just looking for trouble. He had been using his new CB mobile on the 2-hour trip to town and when he overheard some truckers talking about the beavers on Channel 13, he gave his tuner a twirl and listened in. Sure enough, there are several legitimate noncommercial "dating services" operating there most weekends. After waiting twenty minutes for a break, Brad found himself being quizzed by a terrific sounding female voice and finally, after what seemed like forever, he was told to switch over to Channel 5 to "meet" his date for the evening. How it all turned out can best be answered by the fact that Brad and his new wife were reported seen at the annual CB Jamboree of the Gold Coast Patrol in Dixon, California, last year.

March 1976
BAYER
GRAND PRIZE

9.

Tuning in the Future: Is It Here Already?

KER ONE-N
LD SMOKY
IN THIS LIN
S. D. H. & P. T.

Space Age Technology Delivers CB into the Hands of the People

The explosive growth of CB has created fallout that is only now becoming discernible throughout the land. An increasingly dehumanized society has suddenly discovered a simple and effective way of getting people together again. The tremendous popularity of CB has created literally thousands of small neighborly clubs, coffee breaks, and similar CB organizations. The unselfish efforts of several thousand members of REACT, ALERT, and other CB public assistance organizations have demonstrably reduced both the loss of life and property damage on the highway and during fires, floods, and other community disasters. CB language, songs, and stories are becoming part of our popular culture. Surely Jack Jett's concept of universal, two-way communications is on the verge of being realized, even beyond his wildest expectations.

The future of CB, however, is now. Manufacturers, spurred on by the fact that CB sales are growing faster than any consumer electronic product in history, have begun to take CB seriously. Industry analysts predict that the total sales volume of CB equipment in 1976 will run second only behind television, and well ahead of stereo equipment.

The giants of the industry—RCA, General Electric, Motorola, Sony, and Panasonic—have all plunged headlong into the citizens band market. Their entry this year has created quantum leaps in research and development and the state of CB art will advance during the next few years at an almost unbelievable rate. Manufacturers, however, predict a tremendous shortage of transceivers will continue through most of 1977. This was caused partially by the critical shortage of crystals, that began in late 1975. (Technological help is on the way, however, with the imminent introduction of 50-channel transceivers requiring one or two crystals—compared to the 12-14 needed for most of today's sets.)

The FCC, moving with characteristic, if deliberate speed, has doubled the number of available Class-D channels in the 27 MHz range and has announced the possibility of opening up as many as fifty more by the end of 1977. Beyond that lies the prospect of shifting Class-D CB out of high frequency altogether and into VHF or UHF wavelengths, thereby creating literally hundreds of potential new channels. This last suggestion has taken the form of an official proposal by the Electronics Industry Association, which calls for the creation of Class-E Citizens Radio Service, just below the present 220 MHz ham band. Another possibility would be the integration of CBers into the amateur ranks under a so-called "Communicators' License." Either way, the mode of modulation would be narrow-band FM, potentially the finest communications system around. Selective calling, which activates a receiver only when a specially-coded tone is transmitted by the sending station, is already possible and several select call systems will be introduced into the market this year. Adding voice-activated transmit switching is the next step, which will be followed quickly by "duplex" transceivers (double system sets

The shape of the future is bug-like and small. This new integrated circuit developed by Hughes Aircraft can synthesize more than 1000 different channels should FCC regulations change. Right now it can be used in 23 and 50 channel sets, doing $50 to $200 worth of tuning for under $10.

which will allow simultaneous sending and receiving, like ordinary telephones). "Repeater" stations, which pick up CB signals and retransmit them, may eventually extend the range beyond the present maximum of 150 miles (although the FCC will have to approve this move, if it's to be legal).

Solid state electronics will cause the size of today's sets, already about one-quarter the size of the old, tube-type models of the 1960s, to shrink even further. Cigarette-box sized, full-channel receivers are already on the drawing boards of several manufacturers, and one Japanese firm has begun displaying at trade shows a prototype of a citizens band two-way wrist radio more sophisticated than anything Dick Tracy ever dreamed of. Antenna sizes will shrink as well, and at the same time, CB ears will become more selective and more powerful.

Automobile industry sources predict that CB transceivers could become standard factory equipment in most automobiles by the early 1980s—certainly the safety factor inherent in mobile units would provide a strong argument in favor of CB (units are already available as standard add-ons from both Chrysler and General Motors). Officials of both the Teamsters and the American Trucking Association say that CB will soon be mandatory in all over-the-road rigs above a certain size; virtually every independent owner-operator already has his ears.

Even with the expansion of CB to other wavelengths, thoughtless CBers will continue to hog the airways and generally ignore the FCC rules. Automatic transmitter identification by coded (inaudible) signals will make it a lot easier for Uncle Charlie to police the airwaves.

After nearly 18 years, the public has begun to recognize the real potential of CB. A minority of die-hard ham operators and a nearly disbelieving industry to the contrary, CB radio has finally been delivered into the hands of the people. That was the hope that its proponents held for the original Citizens Radio Service and its future now is as unlimited as man's own inventiveness in communicating with his fellow man.

CB Coats of Arms

The heraldry of CB land is as rich and inventive as the language CBers have created for themselves. Full of native funk and flash, the patches worn by each of the literally thousands of CB clubs and coffee breaks are strikingly original interpretations of the unique character and personality of each organization. These patches are worn on CB vests and jackets and traded for other club emblems at CB conventions and jamborees.

ARIZONA
THUNDERBIRD
CB
RADIO CLUB
GLENDALE

WHITE COUNTY EMERGENCY
ARK.
Direct Support For
The Needs Of The
Community
C-B- CLUB

CITIZEN'S BAND
OMAHA
NEBRASKA
COMMUNICATORS

CENTRAL NEBRASKA PEACE MAKERS
19
2
23
NOT C.B.ERS
7
14
33 85
KEARNEY, NEBR.

QUEEN CITY C.B.'ERS
POUGHKEEPSIE, N.Y.

HAPPY VALLEY CB CLUB

WALKING TALKING CB'RS
73'S
88'S
PLENTYWOOD MONTANA

HELPING HAND RADIO CLUB
C
B
LANC. CO. PA.

LAKEVIEW
CIRCUIT BREAKERS
"WE'RE A
BREAKIN
FOR YA'

THE
MIKE BENDERS
C.B. CLUB
BREAK
EAST CENTRAL
MISSOURI

SOUTHWESTERN
HELPING
HAND
MONITOR
23
CB'ERS
PENNA.

UNITED STATES
MEMBER
C.B. CLUB

COTTON-PICKIN'-BANDITS
C.B.-RADIO-CLUB

BAUMHOLDER AMERICAN
GERMANY
RADIO ASSOCIATION

OUTAGAMIE
COUNTY
CB
RADIO ASSIST

REPRESENTATIVE
VALLEY BREAK CONTROL
CENTRAL CAL

CITIZENS BAND
QSL
STREAK CLUB

LORAINE C.B.'ers
LORAINE, TEXAS
JACKSON COUNTY
Ch. 10
CBERS
DON'T RUN BAREFOOT GET A C.B.
19
2
23
THE NATIONS #1 C.B.'ERS CLUB
CENTRAL NEBRASKA
PEACE MAKER'S
KEARNEY, NEBR.
3's
8's
7
14

Glossary of Technical Terms

AF abbreviation for Audio Frequency, the portion of the electromagnetic spectrum that includes sounds that can be heard by humans, generally considered to be 20 HZ-20 HZ (kilohertz).

AM abbreviation for Amplitude Modulation, the type of transmission used for CB transmissions (and regular AM radio) where the strength of a carrier wave is changed in proportion to the changes in the voice being transmitted.

ANL abbreviation for Automatic Noise Limiter, a feature on most CB sets that eliminates most interference caused by auto ignition systems.

ANTENNA MATCHER a device used to compensate for a high standing wave ratio (SWR).

BAND a range of frequencies between two limits. The AM broadcast band extends from 535 to 1605 KHZ; the original Class-D Citizens Band goes from 26.965 to 27.255 MHZ (megahertz).

BEAM a type of antenna that focuses the signal into a small area, increasing range by making use of power that would have otherwise been wasted by being radiated in all directions.

BLEED OVER to interfere with conversations on a channel other than the one you are transmitting on, caused by an improperly operating transmitter, or a transmitter too close to a receiver.

CARRIER the radio wave that is altered (modulated) by changes in amplitude, so it can carry a voice signal. Someone who transmits an unmodulated carrier (holds the mike button without speaking) interrupts other conversations, and is said to be "throwing a carrier."

CHANNEL a specific frequency used in communication, such as TV or CB. The emergency CB channel is Channel 9, and has a frequency of 27.065 MHz.

CLARIFIER a device for fine-tuning single-sideband receivers.

COAXIAL CABLE "coax" for short, is a type of electrical cable with a center conductor surrounded by an insulator which is in turn surrounded by another conductor. The outer conductor acts as a shield to protect the center conductor from electrical interference. In CB, RG/58U (about 10¢/ft) coax is used for runs of up to 50 ft, and heavier RG8/U (about 20¢/ft) for longer runs.

COMPRESSOR device that both amplifies and limits the signal being fed to a transmitter from a microphone to insure maximum modulation without distortion.

COPHASING the use of two antennas connected together to get an increase in power in a particular direction at the expense of another direction.

CRYSTAL a material, such as quartz, that vibrates at a particular frequency when electrified. It is used to set the frequency at which a transmitter or receiver operates.

CYCLE in electronics, the generation of one complete wave, with full strength in both positive and negative directions. One cycle per second, or 1 Hz, is the basic unit of measuring frequency. In CB, there are roughly 27 million complete cycles each second.

dB abbreviation for decibel, the standard unit used to compare audio or radio signal strength. It is a logorithmic unit, and a 3 dB increase represents about twice the power, 6 dB about a 4 times increase, 10 dB a 10 times increase, 20 dB a 100 times increase.

DELTA TUNE a device designed to permit a receiver to tune slightly above or below the desired frequency, useful if the person you want to hear is operating off frequency, or if you want to "get away from" someone on another channel who is bleeding over.

DIRECTIONAL refers to an antenna that is designed to transmit and receive in one specific direction, and must be rotated for conversations with stations in other areas.

DIRECTIONAL ANTENNA see "beam."

DUAL CONVERSION type of receiver circuit that provides particularly good *selectivity*, the rejection of frequencies other than the desired one.

ELECTROMAGNETIC WAVE a wave consisting of electric and magnetic lines of force at right angles to each other, such as radio waves. With a horizontally polarized antenna, the magnetic waves move vertically and the electric waves move horizontally. With a vertical antenna as generally used in CB, the directions are reversed.

ERP abbreviation for Effective Radiated Power, refers to the effect of a gain antenna on a transmitter. A strong antenna can make a 4-watt CB station have an ERP of hundreds of watts, and naturally reach much farther than with a standard antenna.

FREQUENCY in radio, the number of waves generated in a particular time, represented by KHZ or MHz which means thousand (or million) wave cycles per second. CB radio operates at about 27 MHz.

FREQUENCY SYNTHESIZER a circuit that uses several crystals together to generate different frequencies. Without such synthesis, a 23 channel CB set would require 46 crystals, one for each channel on both transmit and receive. A frequency convential synthesizer cuts that number to 14. Advanced tranceivers with digital frequency synthesizers use only one or two crystals.

FRONT-TO-BACK RATIO refers to a beam antenna's ability to reject a signal coming from behind when you want to talk to someone in front of you. This is expressed in dB, usually in the 10–30 dB range, with bigger numbers better than small.

GAIN an increase in power. In CB you can get gain by using a microphone preamplifier, an illegal linear amplifier, or a powerful antenna.

GROUND in CB, either the connection to the chassis of a car that goes to the negative (–) side of the battery; or the earth or a substitute for the earth that serves as part of a base-station antenna.

GROUND PLANE ANTENNA an antenna that uses three or four horizontal or down-tilted radials to take the place of the "earth" ground so it can be mounted in the air; or a mobile antenna that uses the car body as a ground.

GROUND WAVE a kind of radio wave, including CB, that follows the surface of the earth.

HANDSET a microphone and earphone mounted in one housing, commonly used for telephones and becoming popular in CB.

HF abbreviation for high frequency, the frequency band between 3 and 30 MHz, which includes the Citizens Band, and other radio services.

HZ abbreviation for Hertz, which means cycles per second, the measurement unit for frequency.

IMPEDANCE the opposition to the flow of alternating current; a characteristic of various electronic devices, such as antennas, cables, transmitters, speakers, microphones, etc., that must be matched for proper operation.

INVERSE SQUARE LAW a principle of physics that applies to radiated energy, such as radio waves. It says that the strength of a received signal decreases in proportion to the square of the distance from the source. If the distance from a receiver to a transmitter is multiplied by 3, the strength of the received signal is divided by 9.

IONOSPHERE a layer of electrically charged particles in the atmosphere from about 50 miles to 250 miles above the earth that reflects back radio waves.

ISOTROPIC the theoretical antenna that has no gain, to which other antennas are compared.

LED abbreviation for Light Emitting Diode, a solid state electronic lamp that uses very little power and lasts an extremely long time, commonly used as a pilot light or flashing modulation indicator on CB sets.

LINEAR AMPLIFIER a device used to increase the output of a radio transmitter. Linear amplifiers are legal for ham radio use, not legal for CB. Any CBer who is found to *have* a linear and does not also have a ham license will be automatically considered guilty of *using* the linear, and the FCC will yank has license and give him a stiff fine. It's called a *linear* amplifier because the output is directly proportional to the input.

LOADING COIL — a device used to compensate for an antenna being shorter than the proper length for a particular frequency, consisting of a coil of wire attached to the antenna.

MICROPHONE — a device used to change sound waves, such as from the human voice, into electrical signals that can be fed to an amplifier or transmitter.

MODULATION — process of impressing voice information on a carrier wave. One hundred per cent modulation is the theoretical best, and can be approached by loud talkers or people using mike preamplifiers.

NOISE BLANKER — a circuit that momentarily (without being noticed) silences a receiver during short bursts of interference; generally considered better than ANL, but varies considerably from transceiver to transceiver and from car to car.

OMNIDIRECTIONAL ANTENNA — an antenna that transmits and receives with equal strength in all directions.

OVERMODULATION — distorted, harsh, crackly sound caused by speaking too loudly into a microphone or setting a microphone preamplifier for too much gain.

PA — abbreviation for Public Address, a feature on many CB sets that lets you use the unit as an amplifier to talk through a speaker, perhaps mounted under the hood of the car or up on the roof.

PEP — abbreviation for Peak Envelope Power, the maximum power put out by a single sideband transmitter, legally limited to 12 watts.

PHASE-LOCKED LOOP — a circuit used in advanced CB transceivers that provides extreme frequency stability, and uses one crystal to generate all frequencies.

PIEZOELECTRIC — "pressure electricity," refers to the ability of certain materials, such as quartz crystals, to generate electricity when pressure is applied, and to move or vibrate when electricity is applied.

POLARITY — two meanings in CB: (1) the positiveness or negativeness of a power connection point. (2) the direction in which radio waves move; vertically polarized waves, such as CB, move perpendicular to the surface of the earth; horizontally polarized waves, such as TV, move parallel to the earth's surface.

POWER SUPPLY — a device or circuit that supplies electrical power to another device or circuit. Most CB transceivers contain their own power supplies. You can get an "AC power supply" to change 110 volts AC to 12 volts DC when you want to use a mobile CB in your home.

PREAMPLIFIER — in CB, either a device to boost the power of your voice after it is picked up by the microphone and before it is fed to the transmitter; or a device used to boost incoming radio signals before they are fed to the receiver.

PTT — abbreviation for Press (or Push) To Talk. .Refers to the switch on a microphone.

RADIAL — horizontal or downward-pointing antenna element fastened to the base of the vertical element.

RADIO — communication using electromagnetic waves, usually moving through air.

RECEIVER — a device used to pick up radio signals transmitted at a distant point.

RESONATE — to vibrate along with, or to operate perfectly matched with, something else. An antenna of a particular length resonates at only one frequency, such as a CB channel, and must be adjusted for optimum performance on any other frequency.

RF — abbreviation for Radio Frequency, that portion of the electromagnetic spectrum that falls between sound and infrared light, roughly 20 kHz to 3000 Hz.

ROTATOR — a motor with a remote control switch used to turn a beam antenna from one direction to another.

RUBBER DUCKIE short flexible antenna commonly used on police walkie-talkies and recently adapted for CB use.

SELECTIVE CALL a feature on some CBs whereby they are kept quiet until activated by a special tone "code" transmitted by another transceiver; commonly used in business applications.

SHORT WAVE a group of RF frequencies above the standard AM broadcast band, including HF, VHF, and UHF.

SIDEBAND a frequency band or space adjacent to the carrier, which actually carries the voice "information." In AM transmissions, there are two sidebands, each carrying identical information.

SINGLE SIDEBAND a more sophisticated means of transmission than AM, in which one sideband and the central carrier are suppressed, thus requiring less frequency room (bandwidth) and having greater range.

SKIP transmitting over long distances by bouncing the radio wave off the ionosphere. Illegal in CB.

SKY WAVE a type of radio wave, sometimes including CB, that goes up to the sky and either dissipates into space or is reflected back to the earth, often hundreds or thousands of miles from the transmitter.

S METER a meter found on CB sets and other radio receivers that indicates the strength of an incoming signal, measured in S-Units, up to S-9; and in dB above S-9. A transmitter has to be very close to give you a signal 20 dB above S-9. S-9 is a great signal. S-5 is typical. A transmitter "hitting you" with only S-2 or 3 is probably too far to hear you and not worth calling unless the channel is extremely quiet.

SQUELCH an electronic circuit that can be adjusted to keep a receiver quiet until a signal of a particular level is received; used to eliminate noise between transmissions.

SWR the ratio of maximum to minimum voltage at various places along the antenna cable. It is an indication of how well an antenna is matched to a transmitter at a particular frequency, with a high ratio indicating that a lot of the power being pushed out by a transmitter is not going out the antenna but is bouncing back through the cable and just lying there, wasting away. 1:1 is the theoretical best SWR, with anything below 1.5:1 considered acceptable.

TRANSCEIVER a transmitter and receiver housed in a single "box" and usually sharing some common parts.

TRANSMITTER a device used to send the human voice or other signals to a distant point by radio.

TVI short for TeleVision Interference. Sound or picture trouble caused on a television when a nearby transmitter is used. It can be cured by using a low pass filter on the CB or a high pass filter on the TV.

UHF abbreviation for Ultra High Frequency, the frequency band between 300 and 3,000 MHz. It contains TV channels 14 to 83, as well as other radio services.

VHF abbreviation for Very High Frequency, the frequency band between 30 and 300 MHz, which contains television, the FM broadcast band, as well as business and public service frequencies.

VOLT unit of measurement of electromotive force, the force which causes electricity to flow from one point to another.

VOX abbreviation for Voice Operated Switch, a circuit that activates a transmitter when you speak into the microphone and does not require a PTT switch. Convenient, and a safety advantage for mobile CB.

WATT the unit of measurement of electrical power. CB transmitters are limited to 4 watts output, of which maybe 4 hundred-billionths of a watt will reach a nearby receiver.

WAVE a physical activity that rises or falls, or advances and retreats, as it moves through a medium. Ocean waves move through water. Radio waves usually move through the air.

WAVELENGTH the physical size of a radio wave (usually measured in meters) corresponding to the distance the wave travels in one cycle.

YAGI the basic beam antenna which uses several elements to "direct" and "reflect" the transmitted signal to provide gain. Named for its Japanese inventor.

CB Public Assistance Organizations

There are literally hundreds of independent emergency radio teams throughout the country and there's probably one in your community and certainly one in your state. Many CBers, however, prefer to join an existing team or create one of their own that is affiliated with one of the three national organizations, ALERT, REACT, or REST-MARINE.

ALERT and REACT are virtually identical in their purposes, but ALERT will accept individual memberships, while REACT, a nonprofit organization, will only accept team applications. Both organizations supply their members with numerous official materials and a bi-monthly publication.

REST-MARINE is considerably smaller than either ALERT or REACT and concentrates in providing emergency radio service along water routes.

	ALERT (Affiliated League of Emergency Radio Teams) 818 Nat'l Press Bld'g. Washington, D.C. 20004	REACT (Radio Emergency Associated Citizens Teams) 11 E. Wacker Drive Chicago, Illinois 60601	REST-MARINE (Radio Emergency Service Teams) 1039 26th Street South Arlington, Virginia 22202
PURPOSE	Provide emergency road, disaster and community services.	Provide emergency road, disaster and community services. Monitor Channel 9.	To provide emergency marine services.
MEMBERSHIP REQUIREMENTS	Obey all FCC regulations Have CB license (or show proof that an application has been filed).	Only team applications accepted (groups of five or more). Obey all FCC regulations Have CB license (or show proof that an application has been filed). Agree as a team to make every effort to monitor Channel 9 at all times.	Only team applications accepted (groups of five or more). Obey all FCC regulations Have FCC license (or show proof of application).
MEMBERSHIP BENEFITS	Membership materials include membership card, certificate, ALERT badge, decals, stickers, etc. One year subscription to *ALERT 44*, a bi-monthly magazine.	Membership materials include membership card, certificate, decals, stickers etc. One year subscription to *The National REACTer*, a bi-monthly newspaper.	Membership materials include REST card and certificate. One year subscription to *Smoke Signals*, a quarterly newsletter.
ANNUAL DUES	$9.50 ($8.50 for each additional member in the same family).	$10 initial team charter plus $2.00 per year for each member.	$6 annual dues per member.

CB Publications and Bibliography

MAGAZINES

CB Magazine

531 N. Ann Arbor
Oklahoma City, Oklahoma 73127
monthly: $10/year.

This well-established, large-format, major CB periodical has a circulation over 200,000. Editor Leo Sands' publication effectively combines technical product information with comprehensive CB social anecdotes and information.

CB News

2435 Maryland Avenue
Baltimore, Maryland 21218
Monthly: $5/year.

A fiercely independent publication, this monthly is edited from the biscuit burner's point of view: doesn't hesitate to call a *glitch* a *glitch*.

CB News & Truckers Gazette

1545 E. Orange Grove Blvd.
Pasadena, California 91104
Monthly: $4/year.

Friendly, funky, and frankly funny, this digest-sized publication combines CB social life with a little bit of religion and sound technical advice. Definitely one of a kind! Editor/Publisher, "Toad Frog" and his wife "Jabberwocky," are folk heroes on the West Coast of CB land.

CB Whaler

P.O. BOX L-177
New Bedford, Massachusetts 02745
Bi-monthly: $1/copy.

A fascinating and unusual magazine which combines rescue stories, social notes and general CB information in a pleasant format.

Overdrive

P.O. Box 54078
Los Angeles, California 90054
Monthly: $19/year

While not strictly a CB publication, this crusading voice of the independent truckers has an excellent CB section in each large format, great graphics issue.

S-9, The Magazine of Two-Way Radio

14 Vanderventer Avenue
Port Washington, New York 11050
Monthly: $10/year.

One of the oldest and best CB publications available anywhere. Newly enlarged format provides even more space for the kind of comprehensive coverage Editor Tom Kneitel has been providing for years.

SSB News

P.O. BOX 6411
Daytona Beach, Florida 32022
Monthly: $5/year.

Written and edited by a former radio retailer, this specialized monthly is definitely for sidewinders only.

World of CB

12301 Wilshire Blvd.
Los Angeles, California 90025
Monthly: $1/copy.

A new monthly slick which offers little new information, but is a good, cheap product guide to basic equipment.

CB Publications and Bibliography

NEWSPAPERS

CB Trader
9906 Rosedale Highway
Bakersfield, California 93308
Monthly: Free.

A curious, maverick monthly with heavy trucker orientation. Spotty distribution, but often worth looking for.

CB World News
P.O. BOX 1747
Ft. Worth, Texas
Monthly: Free.

This illustrated tabloid is an excellent example of a real CB publication. Put together by two young, enthusiastic CBers, the "World News" covers just about the entire CB world in quick, easy to read fashion.

CBer's News
Columbia, Missouri 65201
Monthly: $5/year.

The official publication of the American CB Association, this tabloid strives to "keep abreast of America's fastest growing past-time."

National CB News
P.O. BOX 28911
Atlanta, Georgia 30328
Monthly: $4/year.

This brand new monthly naively advertises itself as "the only paper in the nation devoted to citizens band operators." Early issues contained less editorial material than most competitors.

NEWSLETTERS

B.C.B [Brotherhood of CBers] Bulletin
P.O. BOX 461
New Haven, Connecticut 06502
Bi-monthly: $5/6 issues.

This quaint mimeographed newsletter views CBers as "one big family," and is full of CB social news, equipment swaps, classifieds and lots of good intentions.

CB Newsletter International
2435 Maryland Avenue
Baltimore, Maryland 21218

A personable mix of social notes and CB gossip, of particular interest to mid-Atlantic biscuit burners.

CB Views
531 N. Ann Arbor
Oklahoma City, Oklahoma 73127
Monthly: $15/year.

Published by CB magazine, this monthly trade newsletter is an excellent source for new product information.

The Communicator
Suite 1
14 Vanderventer Avenue
Port Washington, New York 11050
Monthly: $12/year.

A trade publication for CB equipment, manufacturers and dealers, this fine trade letter from Cowan Publishing contains latest product information, industry personnel changes, and general market information.

Electronic Flea Market
14 Vanderventer Avenue, Suite 1
Port Washington, New York 11050
Monthly: $12/year.

This monthly newsletter is a good source for used equipment, swaps, and classifieds of interest to CBers and other electronic enthusiasts. Free classified space for subscribers.

Ten-4 Newsletter
427 Birchwood Avenue
Deerfield, Illinois 60015

Irregular publication. Subscription included in $10 annual membership fee in ten-4 International, Inc.

Voice of Sideband International
Box 76
Eugene, Oregon 97402
Monthly: $3.50/year.

A specialized newsletter for single sidebanders only.

National CB Clubs and Organizations

Exclusive of public assistance organizations like REACT, there are today over 20,000 different local, state, and regional CB clubs and organizations. Some are as old as CB itself, but most are informal coffee breaks, neighborhood social groups, and broader, CB-oriented statewide organizations. The vast majority of CB social clubs last only about six months to a year and many were literally "born yesterday." Because of the recent boom, CBers would appear numerous enough now to have a national voice and even, perhaps, a Washington lobby (in view of the need for a constant dialogue between CB operators and the FCC). The closest any existing organizations come, however, to having a nationwide constituency are the following:

American CB Radio Association
Box 1702
Columbia, Missouri 65201

American Federation of CBers
Box 5184
Detroit, Michigan 48235

Bay Area Radio Network
Box 1184
Pittsburg, California 94565

Brotherhood of CBers
Box 461
New Haven, Connecticut

CB Club of America
Suite 1607
52 Vanderbilt Avenue
New York, New York 10017

Citizens Band Society of America, Ltd.
Warwick, Rhode Island

Single Sideband Club
All America
Box 647
South Orange, New Jersey 07079

Skip Talkers of America, Inc.
Box 3251
Evansville, Indiana

Ten-Four International, Inc.
427 Birchwood Avenue
Deerfield, Illinois 60015

The Cross-Country CBers Club
Box 54078
Los Angeles, California 90054

United States CB Club
Washington Crossing, Pennsylvania 18977

United States Citizens Radio Council
3600 Noble Street
Anniston, Alabama 36201

CB Handle Registries

With more than 15,000,000 active CBers in the country (and over half a million rubberbanders joining their ranks every month) the difficulty of finding an original CB handle is escalating into an impossible situation. For several years, handle registries have attempted to make some kind of order out of the chaos of handle duplication. Most registries are actually entrepreneurial organizations which have dreamed up a fairly effective system of separating CBers from yet another $5 or $10. But if you do care about having the date of your CB-christening immortalized, the outfits listed below may be worth the expense. (Note: Some registries don't guarantee anything more than exclusive registration by state or region. Others merely log your handle and issue you a "registration" number, without guaranteeing exclusivity of any kind. Unfortunately, there is no final arbiter. The FCC, of course, does not officially recognize handles, and insists that the only *legal* identification you can give on the air is your station call sign.)

American CB Registry
Box 58
Lemont, Illinois 60439
$10, for lifetime registration of your handle.
Package includes newsletter, plastic ID card and "official" seal.

CB Registry of North America
P.O. Box 4703
Hialeah, Florida 33014
$3, for registration only. Does not publish directory.
"CB Registry is the original handle registry" and they guarantee statewide exclusivity of your handle. They will advise you, however, if someone else has beaten you to your handle.

Communication Co-Op
2250 NW 39 Street
Oklahoma City, Oklahoma 73112
$7, for CB-Handle Directory, includes registration of your own handle, permanent listing in subsequent directories, and wall certificate.

National Bureau of Search and Call Registration
Box 76
International Falls, Minnesota 56649
$5, for registration of your handle, wall certificate and walled ID card.

National Handle Club of America, Inc.
706 East Highlaw
Shawnee, Oklahoma 74801
$10, for lifetime membership.
Members receive wall certificate, bumper sticker, "Choctaw Curse" (anti-theft sticker), car decal, and pocket membership card. Only accepts handles not previously listed.

United CB Handle Registry
Dept CBB 876
Box 1232
Mission, Kansas 66222
$3, for 8½″ × 11″ "gold background" certificate only.
Your certificate number, with your handle, will be kept in a permanent record in the files of United CB Registry. Offer also includes copy of Ten Code, "Q" Code, and "CB Slanguage" card.

Directory of Selected Manufacturers

The following list of CB equipment manufacturers, dealers, distributors and suppliers is representative of the best known and most readily available CB equipment and accessories.

The Antenna Specialists Co.
12435 Euclid Ave.
Cleveland, Ohio 44106

Astatic Cog0m3f3Harbor and Jackson Streets
Conneaut, Ohio 44030

Avanti Research and Development,
Inc.
340 Stewart Ave.
Addison, Illinois 60101

Beltek Corporation of America
17910 S. LaSalle Ave.
Gardena, California 90248

Blue Streak
(see CPD)

Breaker Corp.
1101 Great Southwest Pkwy.
Arlington, Texas 76011

Browning Laboratories, Inc.
1269 Union Ave.
Laconia, New Hampshire 03246

Cobra
(see Dynascan)

Communications Power, Inc.
2407 Charleston Rd.
Mountain View, California 94043

Cornell-Dubilier Electronics
150 Avenue L
Newark, New Jersey 07101

CPD Industries, Inc.
2100 E. Wilshire Ave.
Santa Ana, California 92705

Craig Corp.
921 W. Artesia Blvd.
Compton, California 90220

Cush-Craft Corp.
621 Hayward St.
Manchester, New Hampshire 03103

Dorado CB Products
P.O. Box 43
El Paso, Texas 7995 70040

Dynascan Corp.
1801 W. Belleplaine
Chicago, Illinois 60613

Fanon/Courier Corp.
990 S. Fair Oaks Ave.
Pasadena, California 91105

Gem Marine Products, Inc.
356 S. Blvd.
P.O. Box 1408
Lake City, South Carolina 29560

General Electric
Box 4197
Lynchburg, Virginia 24502

Gold Line Connector, Inc.
25 Van Zant St.
E. Norwalk, Connecticut 06855

Handic of U.S.A., Inc.
14340 N.W. 60th Ave.
Miami Lakes, Florida 33014

Hitachi Sales Corp.
401 W. Artesia Blvd.
Compton, California 90220

Hufco
P.O. Box 357
Provo, Utah 84601

Hustler
(see New-Tronics)

Hy-Gain Electronics Corp.
8601 N.E. Hwy. 6
Lincoln, Nebraska 68505

J.I.L. Corporation of America
737 W. Artesia Blvd.
Compton, California 90220

E.F. Johnson Co.
299 10th Ave., SW
Waseca, Minnesota 56093

Kraco Enterprises, Inc.
2411 N. Santa Fe Ave.
Compton, California 90224

Kris, Inc.
Pioneer Rd.
Cedarburg, Wisconsin 53012

Lafayette Radio Electronics
111 Jericho Turnpike
Syosset, New York 11791

Leader Instruments Corp.
151 Dupont St.
Plainview, New York 11803

Linear Systems, Inc.
220 Airport Blvd.
Watsonville, California 95076

Midland International Corp.
Box 19032
Kansas City, Missouri 64141

Mura Corp.
50 S. Service Rd.
Jericho, New York 11753

New-Tronics Corp.
15800 Commerce Park Drive
Brookpark, Ohio 44142

Pace
(see Pathcom)

P.A.L. Electronics Co.
2962 W. Weldon
Phoenix, Arizona 85017

Palomar Electronics Corp.
665 Opper St.
Escondido, California 92025

Panasonic
1 Panasonic Way
Secaucus, New Jersey 07148

Pathcom, Inc.
24049 S. Frampton Ave.
Harbor City, California 90710

Pearce-Simpson
4701 N.W. 77th Ave.
Miami, Florida 33166

Philmore Manufacturing Co., Inc.
40 Inip Dr.
Inwood, New York 11696

Radio Shack
P.O. Box 2625
Fort Worth, Texas 76101

Realistic
(See Radio Shack)

Recoton Corp.
46-23 Crane St.
Long Island City, New York 11101

Regency Electronics
7707 Records St.
Indianapolis, Indiana 46222

Rohn
Box 2000
Peoria, Illinois 61601

Royce Electronics Corp.
1746 Levee Rd.
N. Kansas City, Missouri 64116

Rystl Electronics Corp.
328 N.W. 170th St.
N. Miami, Florida 33160

SBE
(see Linear Systems)

Shakespeare Electronics
Box 246
Columbia, South Carolina 29202

Sharp Electronics Corp.
10 Keystone Pl.
Paramus, New Jersey 07562

Shure Brothers, Inc.
222 Hartyre Ave.
Evanston, Illinois 60204

Shur-Lok Mfg. Co.
413 N. Main St.
Hutchins, Texas 75141

Siltronix
269 Airport Rd.
Oceanside, California 92054

Sonar
73 Wortman Avenue
Brooklyn, New York 11207

Sony
9 West 57th St.
New York, New York 10019

SST
South Shore Trading Co.
1311 Bellmore Ave.
N. Bellmore, New York

Superex Electronics Corp.
151 Ludlow St.
Yonkers, New York 10705

Teaberry Electronics Corp.
4655 Mass. Ave.
Indianapolis, Indiana 46218

Telco Products Corp.
44 Sea Cliff Ave.
Glen Cove, New York 11452

Telex
9600 Aldrich Ave. S.
Minneapolis, Minnesota 55420

Tenna Corp.
19201 Cranwood Parkway
Cleveland, Ohio 44128

Time Machines, Inc.
29 Falmouth St.
Brooklyn, New York 11235

Tram/Diamond Corp.
Box 187
Lower Bay Rd.
Winnisquam, New Hampshire 03289

Turner
909 17th St. N.E.
Cedar Rapids, Iowa 52402

Wilson Electronics
4288 S. Polaris Ave.
Las Vegas, Nevada 89103

CB Goodies Suppliers

The Rambling Redskin (S. Russell)
Box 564
North Bergen, New Jersey 07047

A dealer and distributor of every manner of CB paraphernalia, the Rambling Redskin has been touring coffee breaks, CB club meetings, jamborees, and conventions continuously for 18 years. He is generally regarded as "the best-known single individual in CB," and has supplied patches, badges, emblems, T-shirts, decals, QSL cards, and other original CB items to thousands of CBers across the country. His C-Bikini is a classic.

A.T. Patch Company
Bethlehem Road
Littleton, New Hampshire 03561

Though not strictly in the CB goodies business, this fine organization has designed and manufactured patches, emblems, and badges for literally hundreds of CB clubs and organizations. President Arthur Paradise is an enthusiastic CB supporter and the repository of more CB stories than probably anyone else in the world.

Sirk Shirts
Box 80498
Lincoln, Nebraska

Tom Shelly, "The Nebraska Sweet Pea," manufactures the most extensive line anywhere of exclusive CB T-shirts.

LePhCo.
2860 Pinkerton Road
Zanesville, Ohio 43701

A centrally located, diversified supplier of CB jackets, vests, T-shirts, and embroidered patches. President Lee Heltzer is also an avid CBer and is a member of the Muskingum Valley CB Radio Association.

RULES AND REGULATIONS

Part 95 | *Citizens Radio Service*

FEDERAL COMMUNICATIONS COMMISSION

WITH LATEST CHANGES

Contents—Part 95

AUTHORITY: §§ 95.1 to 95.147 issued under secs. 4, 303, 48 Stat. 1066, 1082, as amended; 47 U.S.C. 154, 303. Interpret or apply 48 Stat. 1064–1068, 1081–1105, as amended; 47 U.S.C. Sub-chap. I, III–VI.

SUBPART A—GENERAL

§ 95.1 Basis and purpose.

The rules and regulations set forth in this part are issued pursuant to the provisions of Title III of the Communications Act of 1934, as amended, which vests authority in the Federal Communications Commission to regulate radio transmissions and to issue licenses for radio stations. These rules are designed to provide for private short-distance radiocommunications service for the business or personal activities of licensees, for radio signaling, for the control of remote objects or devices by means of radio; all to the extent that these uses are not specifically prohibited in this part. They also provide for procedures whereby manufacturers of radio equipment to be used or operated in the Citizens Radio Service may obtain type acceptance and/or type approval of such equipment as may be appropriate.

§ 95.3 Definitions.

For the purpose of this part, the following definitions shall be applicable. For other definitions, refer to Part 2 of this chapter.

(a) Definitions of services.

Citizens Radio Service. A radiocommunications service of fixed, land, and mobile stations intended for short-distance personal or business radiocommunications, radio signaling, and control of remote objects or devices by radio; all to the extent that these uses are not specifically prohibited in this part.

Fixed service. A service of radiocommunication between specified fixed points.

Mobile service. A service of radiocommunication between mobile and land stations or between mobile stations.

(b) Definitions of stations.

Base station. A land station in the land mobile service carrying on a service with land mobile stations.

Class A station. A station in the Citizens Radio Service licensed to be operated on an assigned frequency in the 460–470 MHz band with a transmitter output power of not more than 50 watts.

Class B station. (All operations terminated as of November 1, 1971.)

Class C station. A station in the Citizens Radio Service licensed to be operated on an authorized frequency in the 26.96–27.23 MHz band, or on the frequency 27.255 MHz, for the control of remote objects or devices by radio, or for the remote actuation of devices which are used solely as a means of attracting attention, or on an authorized frequency in the 72–76 MHz band for the radio control of models used for hobby purposes only.

Class D station. A station in the Citizens Radio Service licensed to be operated for radiotelephony, only, on an authorized frequency in the 26.96–27.23 MHz band and on the frequency 27.255 MHz.

Fixed station. A station in the fixed service.

Land station. A station in the mobile service not intended for operation while in motion. (Of the various types of land stations, only the base station is pertinent to this part.)

Mobile station. A station in the mobile service intended to be used while in motion or during halts at unspecified points. (For the purposes of this part, the term includes hand-carried and pack-carried units.)

(c) Miscellaneous definitions.

Antenna structures. The term "antenna structures" includes the radiating system, its supporting structures and any appurtenances mounted thereon.

Assigned frequency. The frequency appearing on a station authorization from which the carrier frequency may deviate by an amount not to exceed that permitted by the frequency tolerance.

Authorized bandwidth. The maximum permissible bandwidth for the particular emission used. This shall be the occupied bandwidth or necessary bandwidth, whichever is greater.

Carrier power. The average power at the output terminals of a transmitter (other than a transmitter having a suppressed, reduced or controlled carrier) during one radio frequency cycle under conditions of no modulation.

Control point. A control point is an operating position which is under the control and supervision of the licensee, at which a person immediately responsible for the proper operation of the transmitter is stationed, and at which adequate means are available to aurally monitor all transmissions and to render the transmitter inoperative.

Dispatch point. A dispatch point is any position from which messages may be transmitted under the supervision of the person at a control point.

Double sideband emission. An emission in which both upper and lower sidebands resulting from the modulation of a particular carrier are transmitted. The carrier, or a portion thereof, also may be present in the emission.

External radio frequency power amplifiers. As defined in § 2.815(a) and as used in this part, an external radio frequency power amplifier is any device which, (1) when used in conjunction with a radio transmitter as a signal source is capable of amplification of that signal, and (2) is not an integral part of a radio transmitter as manufactured.

Harmful interference. Any emission, radiation or induction which endangers the functioning of a radionavigation service or other safety service or seriously degrades, obstructs or repeatedly interrupts a radiocommunication service operating in accordance with applicable laws, treaties, and regulations.

Man-made structure. Any construction other than a tower, mast or pole.

Mean power. The power at the output terminals of a transmitter during normal operation, averaged over a time sufficiently long compared with the period of the lowest frequency encountered in the modulation. A time of 1/10 second during which the mean power is greatest will be selected normally.

Necessary bandwidth. For a given class of emission, the minimum value of the occupied bandwidth sufficient to ensure the transmission of information at the rate and with the quality required for the system employed, under specified conditions. Emissions useful for the good functioning of the receiving equipment, as for example, the emission corresponding to the carrier of reduced carrier systems, shall be included in the necessary bandwidth.

Occupied bandwidth. The frequency bandwidth such that, below its lower and above its upper frequency limits, the mean powers radiated are each equal to 0.5% of the total mean power radiated by a given emission.

Omnidirectional antenna. An antenna designed so the maximum radiation in any horizontal direction is within 3 dB of the minimum radiation in any horizontal direction.

Peak envelope power. The average power at the output terminals of a transmitter during one radio frequency cycle at the highest crest of the modulation envelope, taken under conditions of normal operation.

Person. The term "person" includes an individual, partnership, association, joint-stock company, trust or corporation.

Remote control. The term "remote control" when applied to the use or operation of a citizens radio station means control of the transmitting equipment of that station from any place other than the location of the transmitting equipment, except that direct mechanical control or direct electrical control by wired connections of transmitting equipment from some other point on the same premises, craft or vehicle shall not be considered to be remote control.

Single sideband emission. An emission in which only one sideband is transmitted. The carrier, or a portion thereof, also may be present in the emission.

Station authorization. Any construction permit, license, or special temporary authorization issued by the Commission.

§ 95.5 Policy governing the assignment of frequencies.

(a) The frequencies which may be assigned to Class A stations in the Citizens Radio Service, and the frequencies which are available for use by Class C or Class D stations are listed in Subpart C of this part. Each frequency available for assignment to, or use by, stations in this service is available on a shared basis only, and will not be assigned for the exclusive use of any one applicant; however, the use of a particular frequency may be restricted to (or in) one or more specified geographical areas.

(b) In no case will more than one frequency be assigned to Class A stations for the use of a single applicant in any given area until it has been demonstrated conclusively to the Commission that the assignment of an additional frequency is essential to the operation proposed.

(c) All applicants and licensees in this service shall cooperate in the selection and use of the frequencies assigned or authorized, in order to minimize interference and thereby obtain the most effective use of the authorized facilities.

(d) Simultaneous operation on more than one frequency in the 72–76 MHz band by a transmitter or transmitters of a single licensee is prohibited whenever such operation will cause harmful interference to the operation of other licensees in this service.

§ 95.6 Types of operation authorized.

(a) Class A stations may be authorized as mobile stations, as base stations, as fixed stations, or as base or fixed stations to be operated at unspecified or temporary locations.

(b) Class C and Class D stations are authorized as mobile stations only; however, they may be operated at fixed locations in accordance with other provisions of this part.

§ 95.7 General citizenship requirements.

A station license shall not be granted to or held by a foreign government or a representative thereof.

[*§ 95.7 revised eff. 2-5-75; VI(75)-1*]

SUBPART B—APPLICATIONS AND LICENSES

§ 95.11 Station authorization required.

No radio station shall be operated in the Citizens Radio Service except under and in accordance with an authorization granted by the Federal Communications Commission.

§ 95.13 Eligibility for station license.

(a) Subject to the general restrictions of § 95.7, any person is eligible to hold an authorization to operate a station in the Citizens Radio Service: *Provided*, That if an applicant for a Class A or Class D station authorization is an individual or partnership, such individual or each partner is eighteen or more years of age; or if an applicant for a Class C station authorization is an individual or partnership, such individual or each partner is twelve or more years of age. An unincorporated association, when licensed under the provisions of this paragraph, may upon specific prior approval of the Commission provide radiocommunications for its members.

NOTE: While the basis of eligibility in this service includes any state, territorial, or local governmental entity, or any agency operating by the authority of such governmental entity, including any duly authorized state, territorial, or local civil defense agency, it should be noted that the frequencies available to stations in this service are shared without distinction between all licensees and that no protection is afforded to the communications of any station in this service from interference which may be caused by the authorized operation of other licensed stations.

(b) [Reserved]

(c) No person shall hold more than one Class C and one Class D station license.

§95.14 Mailing address furnished by licensee.

Each application shall set forth and each licensee shall furnish the Commission with an address in the United States to be used by the Commission in serving documents or directing correspondence to that licensee. Unless any licensee advises the Commission to the contrary, the address contained in the licensee's most recent application will be used by the Commission for this purpose.

[*§ 95.14 added new eff. 2-5-75; VI(75)-1*]

§ 95.15 Filing of applications.

(a) To assure that necessary information is supplied in a consistent manner by all persons, standard forms are prescribed for use in connection with the majority of applications and reports submitted for Commission consideration. Standard numbered forms applicable to the Citizens Radio Service are discussed in § 95.19 and may be obtained from the Washington, D.C., 20554, office of the Commission, or from any of its engineering field offices.

(b) All formal applications for Class C or Class D new, modified, or renewal station authorizations shall be submitted to the Commission's office at 334 York Street, Gettysburg, Pa. 17325. Applications for Class A station authorizations, applications for consent to transfer of control of a corporation holding any citizens radio station authorization, requests for special temporary authority or other special requests, and correspondence relating to an application for any class citizens radio station authorization shall be submitted to the Commission's Office at Washington, D.C. 20554, and should be directed to the attention of the Secretary. Beginning January 1, 1973, applicants for Class A stations in the Chicago Regional Area, defined in § 95.19, shall submit their applications to the Commission's Chicago Regional Office. The address of the Regional Office will be announced at a later date. Applications involving Class A or Class D station equipment which is neither type approved nor crystal controlled, whether of commercial or home construction, shall be accompanied by supplemental data describing in detail the design and construction of the transmitter and methods employed in testing it to determine compliance with the technical requirements set forth in Subpart C of this part.

(c) Unless otherwise specified, an application shall be filed at least 60 days prior to the date on which it is desired that Commission action thereon be completed. In any case where the applicant has made timely and sufficient application for renewal of license, in accordance with the Commission's rules, no license with reference to any activity of a continuing nature shall expire until such application shall have been finally determined.

(d) Failure on the part of the applicant to provide all the information required by the application form, or to supply the necessary exhibits or supplementary statements may constitute a defect in the application.

(e) Applicants proposing to construct a radio station on a site located on land under the jurisdiction of the U.S. Forest Service, U.S. Department of Agriculture, or the Bureau of Land Management, U.S. Department of the Interior, must supply the information and must follow the procedure prescribed by § 1.70 of this chapter.

§ 95.17 Who may sign applications.

(a) Except as provided in paragraph (b) of this section, applications, amendments thereto, and related statements of fact required by the Commission shall be personally signed by the applicant, if the applicant is an individual; by one of the partners, if the applicant is a partnership; by an officer, if the applicant is a corporation; or by a member who is an officer, if the applicant is an unincorporated association. Applications, amendments, and related statements of fact filed on behalf of eligible government entities, such as states and territories of the United States and political subdivisions thereof, the District of Columbia, and units of local government, including incorporated municipalities, shall be signed by such duly elected or appointed officials as may be competent to do so under the laws of the applicable jurisdiction.

(b) Applications, amendments thereto, and related statements of fact required by the Commission may be signed by the applicant's attorney in case of the applicant's physical disability or of his absence from the United States. The attorney shall in that event separately set forth the reason why the application is not signed by the applicant. In addition, if any matter is stated on the basis of the attorney's belief only (rather than his knowledge), he shall separately set forth his reasons for believing that such statements are true.

(c) Only the original of applications, amendments, or related statements of fact need be signed; copies may be conformed.

(d) Applications, amendments, and related statements of fact need not be signed under oath. Willful false statements made therein, however, are punishable by fine and imprisonment. U.S. Code, Title 18, section 1001, and by appropriate administrative sanctions, including revocation of station license pursuant to section 312(a)(1) of the Communications Act of 1934, as amended.

§ 95.19 Standard forms to be used.

(a) *FCC Form 505, Application for Class C or D Station License in the Citizens Radio Service.* This form shall be used when:

(1) Application is made for a new Class C or Class D authorization. A separate application shall be submitted for each proposed class of station.

(2) Application is made for modification of any existing Class C or Class D station authorization in those cases where prior Commission approval of certain changes is required (see § 95.35).

(3) Application is made for renewal of an existing Class C or Class D station authorization, or for reinstatement of such an expired authorization.

(b) *FCC Form 400, Application for Radio Station Authorization in the Safety and Special Radio Services.* Except as provided in paragraph (d) of this section, this form shall be used when:

(1) Application is made for a new Class A base station or fixed station authorization. Separate applications shall be submitted for each proposed base or fixed station at different fixed locations; however, all equipment intended to be operated at a single fixed location is considered to be one station which may, if necessary, be classed as both a base station and a fixed station.

(2) Application is made for a new Class A station authorization for any required number of mobile units (including hand-carried and pack-carried units) to be operated as a group in a single radiocommunication system in a particular area. An application for Class A mobile station authorization may be combined with the application for a single Class A base station authorization when such mobile units are to be operated with that base station only.

(3) Application is made for station license of any Class A base station or fixed station upon completion of construction or installation in accordance with the terms and conditions set forth in any construction permit required to be issued for that station, or application for extension of time within which to construct such a station.

(4) Application is made for modification of any existing Class A station authorization in those cases where prior Commission approval of certain changes is required (see § 95.35).

(5) Application is made for renewal of an existing Class A station authorization, or for reinstatement of such an expired authorization.

(6) Each applicant in the Safety and Special Radio Services (1) for modification of a station license involving a site change or a substantial increase in tower height or (2) for a license for a new station must, before commencing construction, supply the environmental information, where required, and must follow the procedure prescribed by Subpart I of Part 1 of this chapter (§§ 1.1301 through 1.1319) unless Commission action authorizing such construction would be a minor action with the meaning of Subpart I of Part 1.

(7) Application is made for an authorization for a new Class A base or fixed station to be operated at unspecified or temporary locations. When one or more individual transmitters are each intended to be operated as a base station or as a fixed station at unspecified or temporary locations for indeterminate periods, such transmitters may be considered to comprise a single station intended to be operated at temporary locations. The application shall specify the general geographic area within which the operation will be confined. Sufficient data must be submitted to show the need for the proposed area of operation.

(c) *FCC Form 703, Application for Consent to Transfer of Control of Corporation Holding Construction Permit or Station License.* This form shall be used when application is made for consent to transfer control of a corporation holding any citizens radio station authorization.

(d) Beginning April 1, 1972, FCC Form 425 shall be used in lieu of FCC Form 400, applicants for Class A stations located in the Chicago Regional Area defined to consist of the counties listed below:

ILLINOIS

1. Boone.
2. Bureau.
3. Carroll.
4. Champaign.
5. Christian.
6. Clark.
7. Coles.
8. Cook.
9. Cumberland.
10. De Kalb.
11. De Witt.
12. Douglas.
13. Du Page.
14. Edgar.
15. Ford.
16. Fulton.
17. Grundy.
18. Henry.
19. Iroquois.
20. Jo Daviess.
21. Kane.
22. Kankakee.
23. Kendall.
24. Knox.
25. Lake.
26. La Salle.
27. Lee.
28. Livingston.
29. Logan.
30. Macon.
31. Marshall.
32. Mason.
33. McHenry.
34. McLean.
35. Menard.
36. Mercer.
37. Moultrie.
38. Ogle.
39. Peoria.
40. Piatt.
41. Putnam.
42. Rock Island.
43. Sangamon.
44. Shelby.
45. Stark.
46. Stephenson.
47. Tazewell.
48. Vermilion.
49. Warren.
50. Whiteside.
51. Will.
52. Winnebago.
53. Woodford.

INDIANA

1. Adams.
2. Allen.
3. Benton.
4. Blackford.
5. Boone.
6. Carroll.
7. Cass.
8. Clay.
9. Clinton.
10. De Kalb.
11. Delaware.
12. Elkhart.
13. Fountain.
14. Fulton.
15. Grant.
16. Hamilton.
17. Hancock.
18. Hendricks.
19. Henry.
20. Howard.
21. Huntington.
22. Jasper.
23. Jay.
24. Kosciusko.
25. Lake.
26. Lagrange.
27. La Porte.
28. Madison.
29. Marion.
30. Marshall.

INDIANA—Continued

31. Miami.
32. Montgomery.
33. Morgan.
34. Newton.
35. Noble.
36. Owen.
37. Parke.
38. Porter.
39. Pulaski.
40. Putnam.
41. Randolph.
42. St. Joseph.
43. Starke.
44. Steuben.
45. Tippecanoe.
46. Tipton.
47. Vermilion.
48. Vigo.
49. Wabash.
50. Warren.
51. Wells.
52. White.
53. Whitley.

IOWA

1. Cedar.
2. Clinton.
3. Dubuque.
4. Jackson.
5. Jones.
6. Muscatine.
7. Scott.

MICHIGAN

1. Allegan.
2. Barry.
3. Berrien.
4. Branch.
5. Calhoun.
6. Cass.
7. Clinton.
8. Eaton.
9. Hillsdale.
10. Ingham.
11. Ionia.
12. Jackson.
13. Kalamazoo.
14. Kent.
15. Lake.
16. Mason.
17. Mecosta.
18. Montcalm.
19. Muskegon.
20. Newaygo.
21. Oceana.
22. Ottawa.
23. St. Joseph.
24. Van Buren.

OHIO

1. Defiance.
2. Mercer.
3. Paulding.
4. Van Wert.
5. Williams.

WISCONSIN

1. Adams.
2. Brown.
3. Calumet.
4. Columbia.
5. Dane.
6. Dodge.
7. Door.
8. Fond du Lac.
9. Grant.
10. Green.
11. Green Lake.
12. Iowa.
13. Jefferson.
14. Juneau.
15. Kenosha.
16. Kewaunee.
17. Lafayette.
18. Manitowoc.
19. Marquette.
20. Milwaukee.
21. Outagamie.
22. Ozaukee.
23. Racine.
24. Richland.
25. Rock.
26. Sauk.
27. Sheboygan.
28. Walworth.
29. Washington.
30. Waukesha.
31. Waupaca.
32. Waushara.
33. Winnebago.

§ 95.25 Amendment or dismissal of application.

(a) Any application may be amended upon request of the applicant as a matter of right prior to the time the application is granted or designated for hearing. Each amendment to an application shall be signed and submitted in the same manner and with the same number of copies as required for the original application.

(b) Any application may, upon written request signed by the applicant or his attorney, be dismissed without prejudice as a matter of right prior to the time the application is granted or designated for hearing.

§ 95.27 Transfer of license prohibited.

A station authorization in the Citizens Radio Service may not be transferred or assigned. In lieu of such transfer or assignment, an application for new station authorization shall be filed in each case, and the previous authorization shall be forwarded to the Commission for cancellation.

§ 95.29 Defective applications.

(a) If an applicant is requested by the Commission to file any documents or information not included in the prescribed application form, a failure to comply with such request will constitute a defect in the application.

(b) When an application is considered to be incomplete or defective, such application will be returned to the applicant, unless the Commission may otherwise direct. The reason for return of the applications will be indicated, and if appropriate, necessary additions or corrections will be suggested.

§ 95.31 Partial grant.

Where the Commission, without a hearing, grants an application in part, or with any privileges, terms, or conditions other than those requested, the action of the Commission shall be considered as a grant of such application unless the applicant shall, within 30 days from the date on which such grant is made, or from its effective date if a later date is specified, file with the Commission a written rejection of the grant as made. Upon receipt of such rejection, the Commission will vacate its original action upon the application and, if appropriate, set the application for hearing.

§ 95.33 License term.

Licenses for stations in the Citizens Radio Service will normally be issued for a term of 5 years from the date of original issuance, major modification, or renewal.

§ 95.35 Changes in transmitters and authorized stations.

Authority for certain changes in transmitters and authorized stations must be obtained from the Commission before the changes are made, while other changes do not require prior Commission approval. The following paragraphs of this section describe the conditions under which prior Commission approval is or is not necessary.

(a) Proposed changes which will result in operation inconsistent with any of the terms of the current authorization require that an application for modification of license be submitted to the Commission. Application for modification shall be submitted in the same manner as an application for a new station license, and the licensee shall forward his existing authorization to the Commission for cancellation immediately upon receipt of the superseding authorization. Any of the following changes to authorized stations may be made only upon approval by the Commission:

(1) Increase the overall number of transmitters authorized.

(2) Change the presently authorized location of a Class A fixed or base station or control point.

(3) Move, change the height of, or erect a Class A station antenna structure.

(4) Make any change in the type of emission or any increase in bandwidth of emission or power of a Class A station.

(5) Addition or deletion of control point(s) for an authorized transmitter of a Class A station.

(6) Change or increase the area of operation of a Class A mobile station or a Class A base or fixed station authorized to be operated at temporary locations.

(7) Change the operating frequency of a Class A station.

(b) When the name of a licensee is changed (without changes in the ownership, control, or corporate structure), or when the mailing address of the licensee is changed (without changing the authorized location of the base or fixed Class A station) a formal application for modification of the license is not required. However, the licensee shall notify the Commission promptly of these changes. The notice, which may be in letter form, shall contain the name and address of the licensee as they appear in the Commission's records, the new name and/or address, as the case may be, and the call signs and classes of all radio stations authorized to the licensee under this part. The notice concerning Class C or D radio stations shall be sent to Federal Communications Commission, Gettysburg, Pa. 17325, and a copy shall be maintained with the records of the station. The notice concerning Class A stations shall be sent to (1) Secretary, Federal Communications Commission, Washington, D.C. 20554, and (2) to Engineer in Charge of the Radio District in which the station is located, and a copy shall be maintained with the license of the station until a new license is issued.

(c) Proposed changes which will not depart from any of the terms of the outstanding authorization for the station may be made without prior Commission approval. Included in such changes is the substitution of transmitting equipment at any station, provided that the equipment employed is included in the Commission's "Radio Equipment List," and is listed as acceptable for use in the appropriate class of station in this service. Provided it is crystal-controlled and otherwise complies with the power, frequency tolerance, emission and modulation percentage limitations prescribed, non-type accepted equipment may be substituted at:

(1) Class C stations operated on frequencies in the 26.99–27.26 MHz band;

(2) Class D stations until November 22, 1974.

(d) Transmitting equipment type accepted for use in Class D stations shall not be modified by the user. Changes which are specifically prohibited include:

(1) Internal or external connection or addition of any part, device or accessory not included by the manufacturer with the transmitter for its type acceptance. This shall not prohibit the external connection of antennas or antenna transmission lines, antenna switches, passive networks for coupling transmission lines or antennas to transmitters, or replacement of microphones.

(2) Modification in any way not specified by the transmitter manufacturer and not approved by the Commission.

(3) Replacement of any transmitter part by a part having different electrical characteristics and ratings from that replaced unless such part is specified as a replacement by the transmitter manufacturer.

(4) Substitution or addition of any transmitter oscillator crystal unless the crystal manufacturer or transmitter manufacturer has made an express determination that the crystal type, as installed in the specific transmitter type, will provide that transmitter type with the capability of operating within the frequency tolerance specified in Section 95.45(a).

(5) Addition or substitution of any component, crystal or combination of crystals, or any other alteration to enable transmission on any frequency not authorized for use by the licensee.

(e) Only the manufacturer of the particular unit of equipment type accepted for use in Class D stations may make the permissive changes allowed under the provisions of Part 2 of this chapter for type acceptance. However, the manufacturer shall not make any of the following changes to the transmitter without prior written authorization from the Commission:

(1) Addition of any accessory or device not specified in the application for type acceptance and approved by the Commission in granting said type acceptance.

(2) Addition of any switch, control, or external connection.

(3) Modification to provide capability for an additional number of transmitting frequencies.

§ 95.37 Limitations on antenna structures.

(a) Except as provided in paragraph (b) of this section, an antenna for a Class A station which exceeds the following height limitations may not be erected or used unless notice has been filed with both the FAA on FAA Form 7460–1 and with the Commission on Form 714 or on the license application form, and prior approval by the Commission has been obtained for:

(1) Any construction or alteration of more than 200 feet in height above ground level at its site (§ 17.7(a) of this chapter).

(2) Any construction or alteration of greater height than an imaginary surface extending outward and upward at one of the following slopes (§ 17.7(b) of this chapter):

(i) 100 to 1 for a horizontal distance of 20,000 feet from the nearest point of the nearest runway of each airport with at least one runway more than 3,200 feet in length, excluding heliports, and seaplane bases without specified boundaries, if that airport is either listed in the Airport Directory of the current Airman's Information Manual or is operated by a Federal military agency.

(ii) 50 to 1 for a horizontal distance of 10,000 feet from the nearest point of the nearest runway of each airport with its longest runway no more than 3,200 feet in length, excluding heliports, and seaplane bases without specified boundaries, if that airport is either listed in the Airport Directory or is operated by a Federal military agency.

(iii) 25 to 1 for a horizontal distance of 5,000 feet from the nearest point of the nearest landing and takeoff area of each heliport listed in the Airport Directory or operated by a Federal military agency.

(3) Any construction or alteration on any airport listed in the Airport Directory of the current Airman's Information Manual (§ 17.7(c) of this chapter).

(b) A notification to the Federal Aviation Administration is not required for any of the following construction or alteration of Class A station antenna structures.

(1) Any object that would be shielded by existing structures of a permanent and substantial character or by natural terrain or topographic features of equal or greater height, and would be located in the congested area of a city, town, or settlement where it is evident beyond all reasonable doubt that the structure so shielded will not adversely affect safety in air navigation. Applicants claiming such exemption shall submit a statement with their application to the Commission explaining the basis in detail for their finding (§ 17.14(a) of this chapter).

(2) Any antenna structure of 20 feet or less in height except one that would increase the height of another antenna structure (§17.14(b) of this chapter).

(c) A Class C or Class D station operated at a fixed location shall employ a transmitting antenna which complies with at least one of the following:

(1) The antenna and its supporting structure does not exceed 20 feet in height above ground level; or

(2) The antenna and its supporting structure does not exceed by more than 20 feet the height of any natural formation, tree or man-made structure on which it is mounted; or

NOTE: A man-made structure is any construction other than a tower, mast, or pole.

(3) The antenna is mounted on the transmitting antenna structure of another authorized radio station and does not exceed the height of the antenna supporting structure of the other station; or

(4) The antenna is mounted on and does not exceed the height of the antenna structure otherwise used solely for receiving purposes, which structure itself complies with subparagraph (1) or (2) of this paragraph.

(5) The antenna is omnidirectional and the highest point of the antenna and its supporting structure does not exceed 60 feet above ground level and the highest point also does not exceed one foot in height above the established airport elevation for each 100 feet of horizontal distance from the nearest point of the nearest airport runway.

NOTE: A work sheet will be made available upon request to assist in determining the maximum permissible height of an antenna structure.

(d) Class C stations operated on frequencies in the 72–76 MHz band shall employ a transmitting antenna which complies with all of the following:

(1) The gain of the antenna shall not exceed that of a half-wave dipole;

(2) The antenna shall be immediately attached to, and an integral part of, the transmitter; and

(3) Only vertical polarization shall be used.

(e) Further details as to whether an aeronautical study and/or obstruction marking and lighting may be required, and specifications for obstruction marking and lighting when required, may be obtained from Part 17 of this chapter, "Construction, Marking, and Lighting of Antenna Structures."

(f) Subpart I of Part 1 of this chapter contains procedures implementing the National Environmental Policy Act of 1969. Applications for authorization of the construction of certain classes of communications facilities defined as "major actions" in § 1.305 thereof, are required to be accompanied by specified statements. Generally these classes are:

(1) Antenna towers or supporting structures which exceed 300 feet in height and are not located in areas devoted to heavy industry or to agriculture.

(2) Communications facilities to be located in the following areas:

(i) Facilities which are to be located in an officially designated wilderness area or in an area whose designation as a wilderness is pending consideration;

(ii) Facilities which are to be located in an officially designated wildlife preserve or in an area whose designation as a wildlife preserve is pending consideration;

(iii) Facilities which will affect districts, sites, buildings, structures or objects, significant in American history, architecture, archaeology or culture, which are listed in the National Register of Historic Places or are eligible for listing (see 36 CFR 800.2 (d) and (f) and 800.10); and

(iv) Facilities to be located in areas which are recognized either nationally or locally for their special scenic or recreational value.

(3) Facilities whose construction will involve extensive change in surface features (e.g. wetland fill, deforestation or water diversion).

NOTE: The provisions of this paragraph do not include the mounting of FM, television or other antennas comparable thereto in size on an existing building or antenna tower. The use of existing routes, buildings and towers is an environmentally desirable alternative to the construction of new routes or towers and is encouraged.

If the required statements do not accompany the application, the pertinent facts may be brought to the attention of the Commission by any interested person during the course of the license term and considered de novo by the Commission.

SUBPART C—TECHNICAL REGULATIONS

§ 95.41 Frequencies available.

(a) Frequencies available for assignment to Class A stations:

(1) The following frequencies or frequency pairs are available primarily for assignment to base and

mobile stations. They may also be assigned to fixed stations as follows:

(i) Fixed stations which are used to control base stations of a system may be assigned the frequency assigned to the mobile units associated with the base station. Such fixed stations shall comply with the following requirements if they are located within 75 miles of the center of urbanized areas of 200,000 or more population.

(*a*) If the station is used to control one or more base stations located within 45 degrees of azimuth, a directional antenna having a front-to-back ratio of at least 15 dB shall be used at the fixed station. For other situations where such a directional antenna cannot be used, a cardioid, bidirectional or omnidirectional antenna may be employed. Consistent with reasonable design, the antenna used must, in each case, produce a radiation pattern that provides only the coverage necessary to permit satisfactory control of each base station and limit radiation in other directions to the extent feasible.

(*b*) The strength of the signal of a fixed station controlling a single base station may not exceed the signal strength produced at the antenna terminal of the base receiver by a unit of the associated mobile station, by more than 6 dB. When the station controls more than one base station, the 6 dB control-to-mobile signal difference need be verified at only one of the base station sites. The measurement of the signal strength of the mobile unit must be made when such unit is transmitting from the control station location or, if that is not practical, from a location within one-fourth mile of the control station site.

(*c*) Each application for a control station to be authorized under the provisions of this paragraph shall be accompanied by a statement certifying that the output power of the proposed station transmitter will be adjusted to comply with the foregoing signal level limitation. Records of the measurements used to determine the signal ratio shall be kept with the station records and shall be made available for inspection by Commission personnel upon request.

(*d*) Urbanized areas of 200,000 or more population are defined in the U.S. Census of Population, 1960, Vol. 1, table 23, page 50. The centers of urbanized areas are determined from the Appendix, page 226 of the U.S. Commerce publication "Air Line Distance Between Cities in the United States."

(ii) Fixed stations, other than those used to control base stations, which are located 75 or more miles from the center of an urbanized area of 200,000 or more population. The centers of urbanized areas of 200,000 or more population are listed on page 226 of the Appendix to the U.S. Department of Commerce publication "Air Line Distance Between Cities in the United States." When the fixed station is located 100 miles or less from the center of such an urbanized area, the power output may not exceed 15 watts. All fixed systems are limited to a maximum of two frequencies and must employ directional antennas with a front-to-back ratio of at least 15 dB. For two-frequency systems, separation between transmit-receive frequencies is 5 MHz.

Base and Mobile (MHz)	*Mobile Only (MHz)*
462.550	467.550
462.575	467.575
462.600	467.600
462.625	467.625
462.650	467.650
462.675	467.675
462.700	467.700
462.725	467.725

(2) Conditions governing the operation of stations authorized prior to March 18. 1968:

(i) All base and mobile stations authorized to operate on frequencies other than those listed in subparagraph (1) of this paragraph may continue to operate on those frequencies only until January 1, 1970.

(ii) Fixed stations located 100 or more miles from the center of any urbanized area of 200,000 or more population authorized to operate on frequencies other than those listed in subparagraph (1) of this paragraph will not have to change frequencies provided no interference is caused to the operation of stations in the land mobile service.

(iii) Fixed stations, other than those used to control base stations, located less than 100 miles (75 miles if the transmitter power output does not exceed 15 watts) from the center of any urbanized area of 200,000 or more population must discontinue operation by November 1, 1971. However, any operation after January 1, 1970, must be on frequencies listed in subparagraph (1) of this paragraph.

(iv) Fixed stations, located less than 100 miles from the center of any urbanized area of 200,000 or more population, which are used to control base stations and are authorized to operate on frequencies other than those listed in subparagraph (1) of this paragraph may continue to operate on those frequencies only until January 1, 1970.

(v) All fixed stations must comply with the applicable technical requirements of subparagraph (1) relating to antennas and radiated signal strength of this paragraph by November 1, 1971.

(vi) Notwithstanding the provisions of subdivisions (i) through (v) of this subparagraph, all stations authorized to operate on frequencies between 465.000 and 465.500 MHz and located within 75 miles of the center of the 20 largest urbanized areas of the United States, may continue to operate on these frequencies only until January 1, 1969. An extension to continue operation on such frequencies until January 1, 1970, may be granted to such station licensees on a case by case basis if the Commission finds that continued operation would not be inconsistent with planned usage of the particular frequency for police purposes. The 20 largest urbanized areas can be found in the U.S. Census of Population, 1960, vol. 1, table 23, page 50. The centers of urbanized areas are determined from the appendix, page 226, of the U.S. Commerce publication, "Air Line Distance Between Cities in the United States."

(b) [Reserved]

(c) Class C mobile stations may employ only amplitude tone modulation or on-off keying of the unmodulated carrier, on a shared basis with other stations in the Citizens Radio Service on the frequencies and under the conditions specified in the following tables:

(1) For the control of remote objects or devices by radio, or for the remote actuation of devices which are used solely as a means of attracting attention and subject to no protection from interference due to the operation of industrial, scientific, or medical devices within the 26.96–27.28 MHz band, the following frequencies are available:

(MHz)	(MHz)	(MHz)
26.995	27.095	27.195
27.045	27.145	[1] 27.255

[1] The frequency 27.255 MHz also is shared with stations in other services.

(2) Subject to the conditions that interference will not be caused to the remote control of industrial equipment operating on the same or adjacent frequencies and to the reception of television transmissions on Channels 4 or 5; and that no protection will be afforded from interference due to the operation of fixed and mobile stations in other services assigned to the same or adjacent frequencies in the band, the following frequencies are available solely for the radio remote control of models used for hobby purposes:

(i) For the radio remote control of any model used for hobby purposes:

MHz	MHz	MHz
72.16	72.32	72.96

(ii) For the radio remote control of aircraft models only:

MHz	MHz	MHz
72.08	72.24	72.40
75.64		

(d) The frequencies listed in the following tables are available for use by Class D mobile stations employing radiotelephony only, on a shared basis with other stations in the Citizens Radio Service, and subject to no protection from interference due to the operation of industrial, scientific, or medical devices within the 26.96–27.28 MHz band.

(1) The following frequencies, commonly known as Channels 1 through 8 and 10 through 23, may be used for communications between units of the same station:

MHz	Channel	MHz	Channel
26.965	1	27.115	13
26.975	2	27.125	14
26.985	3	27.135	15
27.005	4	27.155	16
27.015	5	27.165	17
27.025	6	27.175	18
27.035	7	27.185	19
27.055	8	27.205	20
27.075	10	27.215	21
27.085	11	27.225	22
27.105	12	27.255	23

(2) The frequency 27.065 MHz (Channel 9) shall be used solely for:

(i) Emergency communications involving the immediate safety of life of individuals or the immediate protection of property or

(ii) Communications necessary to render assistance to a motorist.

NOTE: A licensee, before using Channel 9, must make a determination that his communication is either or both (a) an emergency communication or (b) is necessary to render assistance to a motorist. To be an emergency communication, the message must have some direct relation to the immediate safety of life or immediate protection of property. If no immediate action is required, it is not an emergency. What may not be an emergency under one set of circumstances may be an emergency under different circumstances. There are many worthwhile public service communications that do not qualify as emergency communications. In the case of motorist assistance, the message must be necessary to assist a particular motorist and not, except in a valid emergency, motorists in general. If the communications are to be lengthy, the exchange should be shifted to another channel, if feasible, after contact is established. No nonemergency or nonmotorist assistance communications are permitted on Channel 9 even for the limited purpose of calling a licensee monitoring a channel to ask him to switch to another channel. Although Channel 9 may be used for marine emergencies, it should not be considered a substitute for the authorized marine distress system. The Coast Guard has stated it will not "participate directly in the Citizens Radio Service by fitting with and/or providing a watch on any Citizens Band Channel. (Coast Guard Commandant Instruction 2302.6.)"

The following are examples of permitted and prohibited types of communications. They are guidelines and are not intended to be all inclusive.

Permitted	Example message
Yes	"A tornado sighted six miles north of town."
No	"This is observation post number 10. No tornados sighted."
Yes	"I am out of gas on Interstate 95."
No	"I am out of gas in my driveway."
Yes	"There is a four-car collision at Exit 10 on the Beltway, send police and ambulance."
No	"Traffic is moving smoothly on the Beltway."
Yes	"Base to Unit 1, the Weather Bureau has just issued a thunderstorm warning. Bring the sailboat into port."
No	"Attention all motorists. The Weather Bureau advises that the snow tomorrow will accumulate 4 to 6 inches."
Yes	"There is a fire in the building on the corner of 6th and Main Streets."
No	"This is Halloween patrol unit number 3. Everything is quiet here."

The following priorities should be observed in the use of Channel 9.

1. Communications relating to an existing situation dangerous to life or property, i.e., fire, automobile accident.

2. Communications relating to a potentially hazardous situation, i.e., car stalled in a dangerous place, lost child, boat out of gas.

3. Road assistance to a disabled vehicle on the highway or street.

4. Road and street directions.

(3) The frequency 27.085 MHz (Channel 11) shall be used only as a calling frequency for the sole purpose of establishing communications and moving to another frequency (channel) to conduct communications.

(e) Upon specific request accompanying application for renewal of station authorization, a Class A station in this service, which was authorized to operate on a frequency in the 460–461 MHz band until March 31, 1967, may be assigned that frequency for continued use until not later than March 31, 1968, subject to all other provisions of this part.

§ 95.43 Transmitter power.

(a) Transmitter power is the power at the transmitter output terminals and delivered to the antenna, antenna transmission line, or any other impedance-matched, radio frequency load.

(1) For single sideband transmitters and other transmitters employing a reduced carrier, a suppressed carrier or a controlled carrier, used at Class D stations, transmitter power is the peak envelope power.

(2) For all transmitters other than those covered by paragraph (a)(1) of this section, the transmitter power is the carrier power.

(b) The transmitter power of a station shall not exceed the following values under any condition of modulation or other circumstances.

Class of station:	Transmitter power in watts
A	50
C—27.255 MHz	25
C—26.995–27.195 MHz	4
C—72–76 MHz	0.75
D—Carrier (where applicable)	4
D—Peak envelope power (where applicable)	12

§ 95.44 External radio frequency power amplifiers prohibited.

No external radio frequency power amplifier shall be used or attached, by connection, coupling attachment or in any other way at any Class D station.

NOTE: An external radio frequency power amplifier at a Class D station will be presumed to have been used where it is in the operator's possession or on his premises and there is extrinsic evidence of any operation of such Class D station in excess of power limitations provided under this rule part unless the operator of such equipment holds a station license in another radio service under which license the use of the said amplifier at its maximum rated output power is permitted.

§ 95.45 Frequency tolerance.

(a) Except as provided in paragraphs (b) and (c) of this section, the carrier frequency of a transmitter in this service shall be maintained within the following percentage of the authorized frequency:

Class of station	Frequency tolerance	
	Fixed and base	Mobile
A	0.00025	0.0005
C		.005
D		.005

(b) Transmitters used at Class C stations operating on authorized frequencies between 26.99 and 27.26 MHz with 2.5 watts or less mean output power, which are used solely for the control of remote objects or devices by radio (other than devices used solely as a means of attracting attention), are permitted a frequency tolerance of 0.01 percent.

(c) Class A stations operated at a fixed location used to control base stations, through use of a mobile only frequency, may operate with a frequency tolerance of 0.0005 percent.

§ 95.47 Types of emission.

(a) Except as provided in paragraph (e) of this section, Class A stations in this service will normally be authorized to transmit radiotelephony only. However, the use of tone signals or signaling devices solely to actuate receiver circuits, such as tone operated squelch or selective calling circuits, the primary function of which is to establish or establish and maintain voice communications, is permitted. The use of tone signals solely to attract attention is prohibited.

(b) [Reserved]

(c) Class C stations in this service are authorized to use amplitude tone modulation or on-off unmodulated carrier only, for the control of remote objects or devices by radio or for the remote actuation of devices which are used solely as a means of attracting attention. The transmission of any form of telegraphy, telephony or record communications by a Class C station is prohibited. Telemetering, except for the transmission of simple, short duration signals indicating the presence or absence of a condition or the occurrence of an event, is also prohibited.

(d) Transmitters used at Class D stations in this service are authorized to use amplitude voice modulation, either single or double sideband. Tone signals or signalling devices may be used only to actuate receiver circuits, such as tone operated squelch or selective calling circuits, the primary function of which is to establish or maintain voice communications. The use of any signals solely to attract attention or for the control of remote objects or devices is prohibited.

(e) Other types of emission not described in paragraph (a) of this section may be authorized for Class A citizens radio stations upon a showing of need therefor. An application requesting such authorization shall fully describe the emission desired, shall indicate the bandwidth required for satisfactory communication, and shall state the purpose for which such emission is required. For information regarding the classification of emissions and the calculation of bandwidth, reference should be made to Part 2 of this chapter.

§ 95.49 Emission limitations.

(a) Each authorization issued to a Class A citizens radio station will show, as a prefix to the classification of the authorized emission, a figure specifying the maximum bandwidth to be occupied by the emission.

(b) [Reserved]

(c) The authorized bandwidth of the emission of any transmitter employing amplitude modulation shall be 8 kHz for double sideband, 4 kHz for single sideband and the authorized bandwidth of the emission of transmitters employing frequency or phase modulation (Class F2 or F3) shall be 20 kHz. The use of Class F2 and F3 emissions in the frequency band 26.96–27.28 MHz is not authorized.

(d) The mean power of emissions shall be attenuated below the mean power of the transmitter in accordance with the following schedule:

(1) When using emissions other than single sideband:

(i) On any frequency removed from the center of the authorized bandwidth by more than 50 percent up to and including 100 percent of the authorized bandwidth: at least 25 decibels;

(ii) On any frequency removed from the center of the authorized bandwidth by more than 100 percent up to and including 250 percent of the authorized bandwidth: At least 35 decibels;

(2) When using single sideband emissions:

(i) On any frequency removed from the center of the authorized bandwidth by more than 50 percent up to and including 150 percent of the authorized bandwidth: At least 25 decibels:

(ii) On any frequency removed from the center of the authorized bandwidth by more than 150 percent up to and including 250 percent of the authorized bandwidth: At least 35 decibels;

(3) On any frequency removed from the center of the authorized bandwidth by more than 250 percent of the authorized bandwidth: At least 43 plus 10 $\log_{10}$ (mean power in watts) decibels.

(e) When an unauthorized emission results in harmful interference, the Commission may, in its discretion, require appropriate technical changes in equipment to alleviate the interference.

§ 95.51 Modulation requirements.

(a) When double sideband, amplitude modulation is used for telephony, the modulation percentage shall be sufficient to provide efficient communication and shall not exceed 100 percent.

(b) Each transmitter for use in Class D stations, other than single sideband, suppressed carrier, or controlled carrier, for which type acceptance is requested after May 24, 1974, having more than 2.5 watts maximum output power shall be equipped with a device which automatically prevents modulation in excess of 100 percent on positive and negative peaks.

(c) The maximum audio frequency required for satisfactory radiotelephone intelligibility for use in this service is considered to be 3000 Hz.

(d) Transmitters for use at Class A stations shall be provided with a device which automatically will prevent greater than normal audio level from causing modulation in excess of that specified in this subpart; *Provided, however*, That the requirements of this paragraph shall not apply to transmitters authorized at mobile stations and having an output power of 2.5 watts or less.

(e) Each transmitter of a Class A station which is equipped with a modulation limiter in accordance with the provisions of paragraph (d) of this section shall also be equipped with an audio low-pass filter. This audio low-pass filter shall be installed between the modulation limiter and the modulated stage and, at audio frequencies between 3 kHz and 20 kHz, shall have an attenuation greater than the attenuation at 1 kHz by at least:

$$60 \log_{10} (f/3) \text{ decibels}$$

where "f" is the audio frequency in kHz. At audio frequencies above 20 kHz, the attentuation shall be at least 50 decibels greater than the attenuation at 1 kHz.

(f) Simultaneous amplitude modulation and frequency or phase modulation of a transmitter is not authorized.

(g) The maximum frequency deviation of frequency modulated transmitters used at Class A stations shall not exceed ±5 kHz.

§ 95.53 Compliance with technical requirements.

(a) Upon receipt of notification from the Commission of a deviation from the technical requirements of the rules in this part, the radiations of the transmitter involved shall be suspended immediately, except for necessary tests and adjustments, and shall not be resumed until such deviation has been corrected.

(b) When any citizens radio station licensee receives a notice of violation indicating that the station has been operated contrary to any of the provisions contained in Subpart C of this part, or where it otherwise appears that operation of a station in this service may not be in accordance with applicable technical standards, the Commission may require the licensee to conduct such tests as may be necessary to determine whether the equipment is capable of meeting these standards and to make such adjustments as may be necessary to assure compliance therewith. A licensee who is notified that he is required to conduct such tests and/or make adjustments must, within the time limit specified in the notice, report to the Commission the results thereof.

(c) All tests and adjustments which may be required in accordance with paragraph (b) of this section shall be made by, or under the immediate supervision of, a person holding a first- or second-class commercial operator license, either radiotelephone or radio telegraph as may be appropriate for the type of emission employed. In each case, the report which is submitted to the Commission shall be signed by the licensed commercial operator. Such report shall describe the results of the tests and adjustments, the test equipment and procedures used, and shall state the type, class, and serial number of the operator's license. A copy of this report shall also be kept with the station records.

§ 95.55 Acceptability of transmitters for licensing.

Transmitters type approved or type accepted for use under this part are included in the Commission's Radio Equipment List. Copies of this list are available for public reference at the Commission's Washington, D.C., offices and field offices. The requirements for transmitters which may be operated under a license in this service are set forth in the following paragraphs.

(a) Class A stations: All transmitters shall be type accepted.

(b) Class C stations:

(1) Transmitters operated in the band 72–76 MHz shall be type accepted.

(2) All transmitters operated in the band 26.99–27.26 MHz shall be type approved, type accepted or crystal controlled.

(c) Class D Stations:

(1) All transmitters first licensed, or marketed as specified in § 2.805 of this chapter, prior to November 22, 1974, shall be type accepted or crystal controlled.

(2) All transmitters first licensed, or marketed as specified in § 2.803 of this chapter, on or after November 22, 1974, shall be type accepted.

(3) Effective November 23, 1978, all transmitters shall be type accepted.

(4) Transmitters which are equipped to operate on any frequency not included in § 95.41(d)(1) may not be installed at, or used by, any Class D station unless there is a station license posted at the transmitter location, or a transmitter identification card (FCC Form 452–C) attached to the transmitter, which indicates that operation of the transmitter on such frequency has been authorized by the Commission.

(d) With the exception of equipment type approved for use at a Class C station, all transmitting equipment authorized in this service shall be crystal controlled.

(e) No controls, switches or other functions which can cause operation in violation of the technical regulations of this part shall be accessible from the operating panel or exterior to the cabinet enclosing a transmitter authorized in this service.

§ 95.57 Procedure for type acceptance of equipment.

(a) Any manufacturer of a transmitter built for use in this service, except noncrystal controlled transmitters for use at Class C stations, may request type acceptance for such transmitter in accordance with the type acceptance requirements of this part, following the type acceptance procedure set forth in Part 2 of this chapter.

(b) Type acceptance for an individual transmitter may also be requested by an applicant for a station authorization by following the type acceptance procedures set forth in Part 2 of this chapter. Such transmitters, if accepted, will not normally be included on the Commission's "Radio Equipment List", but will be individually enumerated on the station authorization.

(c) Additional rules with respect to type acceptance are set forth in Part 2 of this chapter. These rules include information with respect to withdrawal of type acceptance, modification of type-accepted equipment, and limitations on the findings upon which type acceptance is based.

(d) Transmitters equipped with a frequency or frequencies not listed in § 95.41(d)(1) will not be type accepted for use at Class D stations unless the transmitter is also type accepted for use in the service in which the frequency is authorized, if type acceptance in that service is required.

§ 95.58 Additional requirements for type acceptance.

(a) All transmitters shall be crystal controlled.

(b) Except for transmitters type accepted for use at Class A stations, transmitters shall not include any provisions for increasing power to levels in excess of the pertinent limits specified in Section 95.43.

(c) In addition to all other applicable technical requirements set forth in this part, transmitters for which type acceptance is requested after May 24, 1974, for use at Class D stations shall comply with the following:

(1) Single sideband transmitters and other transmitters employing reduced, suppressed or controlled carrier shall include a means for automatically preventing the transmitter power from exceeding either the maximum permissible peak envelope power or the rated peak envelope power of the transmitter, whichever is lower.

(2) Multi-frequency transmitters shall not provide more than 23 transmitting frequencies, and the frequency selector shall be limited to a single control.

(3) Other than the channel selector switch, all transmitting frequency determining circuitry, including crystals, employed in Class D station equipment shall be internal to the equipment and shall not be accessible from the exterior of the equipment cabinet or operating panel.

(4) Single sideband transmitters shall be capable of transmitting on the upper sideband. Capability for transmission also on the lower sideband is permissible.

(5) The total dissipation ratings, established by the manufacturer of the electron tubes or semiconductors which supply radio frequency power to the antenna terminals of the transmitter, shall not exceed 10 watts. For electron tubes, the rating shall be the Intermittent Commercial and Amateur Service (ICAS plate dissipation value if established. For semiconductors, the rating shall be the collector or device dissipation value, whichever is greater, which may be temperature de-rated to not more than 50°C.

(d) Only the following external transmitter controls, connections or devices will normally be permitted in transmitters for which type acceptance is requested after May 24, 1974, for use at Class D stations. Approval of additional controls, connections or devices may be given after consideration of the function to be performed by such additions.

(1) Primary power connection. (Circuitry or devices such as rectifiers, transformers, or inverters which provide the nominal rated transmitter primary supply voltage may be used without voiding the transmitter type acceptance.)

(2) Microphone connection.

(3) Radio frequency output power connection.

(4) Audio frequency power amplifier output connector and selector switch.

(5) On-off switch for primary power to transmitter. May be combined with receiver controls such as the receiver on-off switch and volume control.

(6) Upper-lower sideband selector; for single sideband transmitters only.

(7) Selector for choice of carrier level; for single sideband transmitters only. May be combined with sideband selector.

(8) Transmitting frequency selector switch.

(9) Transmit-receive switch.

(10) Meter(s) and selector switch for monitoring transmitter performance.

(11) Pilot lamp or meter to indicate the presence of radio frequency output power or that transmitter control circuits are activated to transmit.

(e) An instruction book for the user shall be furnished with each transmitter sold and one copy (a draft or preliminary copy is acceptable providing a final copy is furnished when completed) shall be forwarded to the Commission with each request for type acceptance or type approval. The book shall contain all information necessary for the proper installation and operation of the transmitter including:

(1) Instructions concerning all controls, adjustments and switches which may be operated or adjusted without causing violation of technical regulations of this part;

(2) Warnings concerning any adjustment which, according to the rules of this part, may be made only by, or under the immediate supervision of, a person holding a commercial first or second class radio operator license;

(3) Warnings concerning the replacement or substitution of crystals, tubes or other components which could cause violation of the technical regulations of this part and of the type acceptance or type approval requirements of Part 2 of this chapter.

(4) Warnings concerning licensing requirements and details concerning the application procedures for licensing.

§ 95.59 Submission of noncrystal controlled Class C station transmitters for type approval.

Type approval of noncrystal controlled transmitters for use at Class C stations in this service may be requested in accordance with the procedure specified in Part 2 of this chapter.

§ 95.61 Type approval of receiver-transmitter combinations.

Type approval will not be issued for transmitting equipment for operation under this part when such equipment is enclosed in the same cabinet, is constructed on the same chassis in whole or in part, or is identified with a common type or model number with a radio receiver, unless such receiver has been certificated to the Commission as complying with the requirements of Part 15 of this chapter.

§ 95.63 Minimum equipment specifications.

Transmitters submitted for type approval in this service shall be capable of meeting the technical specifications contained in this part, and in addition, shall comply with the following:

(a) Any basic instructions concerning the proper adjustment, use, or operation of the equipment that may be necessary shall be attached to the equipment in a suitable manner and in such positions as to be easily read by the operator.

(b) A durable nameplate shall be mounted on each transmitter showing the name of the manufacturer, the type or model designation, and providing suitable space for permanently displaying the transmitter serial number, FCC type approval number, and the class of station for which approved.

(c) The transmitter shall be designed, constructed, and adjusted by the manufacturer to operate on a frequency or frequencies available to the class of station for which type approval is sought. In designing the equipment, every reasonable precaution shall be taken to protect the user from high voltage shock and radio frequency burns. Connections to batteries (if used) shall be made in such a manner as to permit replacement by the user without causing improper operation of the transmitter. Generally accepted modern engineering principles shall be utilized in the generation of radio frequency currents so as to guard against unnecessary interference to other services. In cases of harmful interference arising from the design, construction, or operation of the equipment, the Commission may require appropriate technical changes in equipment to alleviate interference.

(d) Controls which may effect changes in the carrier frequency of the transmitter shall not be accessible from the exterior of any unit unless such accessibility is specifically approved by the Commission.

§ 95.65 Test procedure.

Type approval tests to determine whether radio equipment meets the technical specifications contained in this part will be conducted under the following conditions:

(a) Gradual ambient temperature variations from 0° to 125° F.

(b) Relative ambient humidity from 20 to 95 percent. This test will normally consist of subjecting the equipment for at least three consecutive periods of 24 hours each, to a relative ambient humidity of 20, 60, and 95 percent, respectively, at a temperature of approximately 80° F.

(c) Movement of transmitter or objects in the immediate vicinity thereof.

(d) Power supply voltage variations normally to be encountered under actual operating conditions.

(e) Additional tests as may be prescribed, if considered necessary or desirable.

§ 95.67 Certificate of type approval.

A certificate or notice of type approval, when issued to the manufacturer of equipment intended to be used or operated in the Citizens Radio Service, constitutes a recognition that on the basis of the test made, the particular type of equipment appears to have the capability of functioning in accordance with the technical specifications and regulations contained in this part: *Provided*, That all such additional equipment of the same type is properly constructed, maintained, and operated: *And provided further*, That no change whatsoever is made in the design or construction of such equipment except upon specific approval by the Commission.

SUBPART D—STATION OPERATING REQUIREMENTS

§95.81 Permissible communications

Stations licensed in the Citizens Radio Service are authorized to transmit the following types of communications:

(a) Communications to facilitate the personal or business activities of the licensee.

(b) Communication relating to:

(1) the immediate safety of life or the immediate protection of property in accordance with §95.85.

(2) the rendering of assistance to a motorist, mariner or other traveler.

(3) civil defense activities in accordance with §95.87.

(4) other activities only as specifically authorized pursuant to §95.87.

(c) Communications with stations authorized in other radio services except as prohibited in §95.83 (a) (3).

§95.83 Prohibited communications

(a) A citizens radio station shall not be used:

(1) For any purpose, or in connection with any activity, which is contrary to Federal, State, or local law.

(2) For the transmission of communications containing obscene, indecent, profane words, language, or meaning.

(3) To communicate with an Amateur Radio Service station, an unlicensed station, or foreign stations (other than as provided in Subpart E of this part) except for communications pursuant to §§95.85 (b) and 95.121.

(4) To convey program material for retransmission, live or delayed, on a broadcast facility.

NOTE: A Class A or Class D station may be used in connection with the administrative, engineering, or maintenance activities of a broadcasting station; a Class A or Class C station may be used for control functions by radio which do not involve the transmission of program material; and a Class A or Class D station may be used in the gathering of news items or preparation of programs: *Provided*, That the actual or recorded transmissions of the Citizens radio station are not broadcast at any time in whole or in part.

(5) To intentionally interfere with the communications of another station.

(6) For the direct transmission of any material to the public through a public address system or similar means.

(7) For the transmission of music, whistling, sound effects, or any material for amusement or entertainment purposes, or solely to attract attention.

(8) To transmit the word "MAYDAY" or other international distress signals, except when the station is located in a ship, aircraft, or other vehicle which is threatened by grave and imminent danger and requests immediate assistance.

(9) For advertising or soliciting the sale of any goods or services.

(10) For transmitting messages in other than plain language. Abbreviations including nationally or internationally recognized operating signals, may be used only if a list of all such abbreviations and their meaning is kept in the station records and made available to any Commission representative on demand.

(11) To carry on communications for hire, whether the renumeration or benefit received is direct or indirect.

(b) A Class D station may not be used to communicate with, or attempt to communicate with, any unit of the same or another station over a distance of more than 150 miles.

(c) A licensee of a Citizens radio station who is engaged in the business of selling Citizens radio transmitting equipment shall not allow a customer to operate under his station license. In addition, all communications by the licensee for the purpose of demonstrating such equipment shall consist only of brief messages addressed to other units of the same station.

§95.85 Emergency and assistance to motorist use.

(a) All Citizens radio stations shall give priority to the emergency communications of other stations which involve the immediate safety of life of individuals or the immediate protection of property.

(b) Any station in this service may be utilized during an emergency involving the immediate safety of life of individuals or the immediate protection of property for the transmission of emergency communications. It may also be used to transmit communications necessary to render assistance to a motorist.

(1) When used for transmission of emergency communications certain provisions in this part concerning use of frequencies (§95.41 (d)); prohibited uses (§95.83 (a) (3); operation by or on behalf of persons other than the licensee (§95.91 (a) and (b)) shall not apply.

(2) When used for transmission of communications necessary to render assistance to a traveler, the provisions of this Part concerning duration of transmission (§95.91 (b)) shall not apply.

(3) The exemptions granted from certain rule provisions in subparagraphs (1) and (2) of this paragraph may be rescinded by the Commission at its discretion.

(c) If the emergency use under paragraph (b) of this section extends over a period of 12 hours or more, notice shall be sent to the Commission in Washington, D.C., as soon as it is evident that the emergency has or will exceed 12 hours. The notice should include the identity of the stations participating, the nature of the emergency, and the use made of the stations. A single notice covering all participating stations may be submitted.

§ 95.87 Operation by, or on behalf of, persons other than the licensee.

(a) Transmitters authorized in this service must be under the control of the licensee at all times. A licensee shall not transfer, assign, or dispose of, in any manner, directly or indirectly, the operating authority under his station license, and shall be responsible for the proper operation of all units of the station.

(b) Citizens radio stations may be operated only by the following persons, except as provided in paragraph (c) of this section:

(1) The licensee;

(2) Members of the licensee's immediate family living in the same household;

(3) The partners, if the licensee is a partnership, provided the communications relate to the business of the partnership;

(4) The members, if the licensee is an unincorporated association, provided the communications relate to the business of the association;

(5) Employees of the licensee only while acting within the scope of their employment;

(6) Any person under the control or supervision of the licensee when the station is used solely for the control of remote objects or devices, other than devices used only as a means of attracting attention; and

(7) Other persons, upon specific prior approval of the Commission shown on or attached to the station license, under the following circumstances:

(i) Licensee is a corporation and proposes to provide private radiocommunication facilities for the transmission of messages or signals by or on behalf of its parent corporation, another subsidiary of the parent corporation, or its own subsidiary. Any remuneration or compensation received by the licensee for the use of the radiocommunication facilities shall be governed by a contract entered into by the parties concerned and the total of the compensation shall not exceed the cost of providing the facilities. Records which show the cost of service and its nonprofit or cost-sharing basis shall be maintained by the licensee.

(ii) Licensee proposes the shared or cooperative use of a Class A station with one or more other licensees in this service for the purpose of communicating on a regular basis with units of their respective Class A stations, or with units of other Class A stations if the communications transmitted are otherwise permissible. The use of these private radiocommunication facilities shall be conducted pursuant to a written contract which shall provide that contributions to capital and operating expense shall be made on a nonprofit, cost-sharing basis, the cost to be divided on an equitable basis among all parties to the agreement. Records which show the cost of service and its nonprofit, cost-sharing basis shall be maintained by the licensee. In any case, however, licensee must show a separate and independent need for the particular units proposed to be shared to fulfill his own communications requirements.

(iii) Other cases where there is a need for other persons to operate a unit of licensee's radio station. Requests for authority may be made either at the time of the filing of the application for station license or thereafter by letter. In either case, the licensee must show the nature of the proposed use and that it relates to an activity of the licensee, how he proposes to maintain control over the transmitters at all times, and why it is not appropriate for such other person to obtain a station license in his own name. The authority, if granted, may be specific with respect to the names of the persons who are permitted to operate, or may authorize operation by unnamed persons for specific purposes. This authority may be revoked by the Commission, in its discretion, at any time.

(c) An individual who was formerly a citizens radio station licensee shall not be permitted to operate any citizens radio station of the same class licensed to another person until such time as he again has been issued a valid radio station license of that class, when his license has been:

(1) Revoked by the Commission.

(2) Surrendered for cancellation after the institution of revocation proceedings by the Commission.

(3) Surrendered for cancellation after a notice of apparent liability to forfeiture has been served by the Commission.

§ 95.89 Telephone answering services.

(a) Notwithstanding the provisions of § 95.87, a licensee may install a transmitting unit of his station on the premises of a telephone answering service. The same unit may not be operated under the authorization of more than one licensee. In all cases, the licensee must enter into a written agreement with the answering service. This agreement must be kept with the licensee's station records and must provide, as a minimum, that:

(1) The licensee will have control over the operation of the radio unit at all times;

(2) The licensee will have full and unrestricted access to the transmitter to enable him to carry out his responsibilities under his license;

(3) Both parties understand that the licensee is fully responsible for the proper operation of the citizens radio station; and

(4) The unit so furnished shall be used only for the transmission of communications to other units belonging to the licensee's station.

(b) A citizens radio station licensed to a telephone answering service shall not be used to relay messages or transmit signals to its customers.

§ 95.91 Duration of transmissions.

(a) All communications or signals, regardless of their nature, shall be restricted to the minimum practicable transmission time. The radiation of energy shall be limited to transmissions modulated or keyed for actual permissible communications, tests, or control signals. Continuous or uninterrupted transmissions from a single station or between a number of communicating stations is prohibited, except for communications involving the immediate safety of life or property.

(b) All communications between Class D stations (interstation) shall be restricted to not longer than five continous minutes. At the conclusion of this 5 minute period, or the exchange of less than 5 minutes, the participating stations shall remain silent for at least one minute.

(c) All communication between units of the same Class D Station (intrastation) shall be restricted to the minimum practicable transmission.

(d) The transmission of audible tone signals or a sequence of tone signals for the operation of the tone operated squelch or selective calling circuits in accordance with § 95.47 shall not exceed a total of 15 seconds duration. Continuous transmission of a subaudible tone for this purpose is permitted. For the purposes of this section, any tone or combination of tones having no frequency above 150 hertz shall be considered subaudible.

(e) The transmission of permissible control signals shall be limited to the minimum practicable time necessary to accomplish the desired control or actuation of remote objects or devices. The continuous radiation of energy for periods exceeding 3 minutes duration for the purpose of transmission of control signals shall be limited to control functions requiring at least one or more changes during each minute of such transmission. However, while it is actually being used to control model aircraft in flight by means of interrupted tone modulation of its carrier, a citizens radio station may transmit a continuous carrier without being simultaneously modulated if the presence or absence of the carrier also performs a control function. An exception to the limitations contained in this paragraph may be authorized upon a satisfactory showing that a continuous control signal is required to perform a control function which is necessary to insure the safety of life or property.

§ 95.93 Tests and adjustments.

All tests or adjustments of citizens radio transmitting equipment involving an external connection to the radio frequency output circuit shall be made using a nonradiating dummy antenna. However, a brief test signal, either with or without modulation, as appropriate, may be transmitted when it is necessary to adjust a transmitter to an antenna for a new station installation or for an existing installation involving a change of antenna or change of transmitters, or when necessary for the detection, measurement, and suppression of harmonic or other spurious radiation. Test transmissions using a radiating antenna shall not exceed a total of 1 minute during any 5-minute period, shall not interfere with communications already in progress on the operating frequency, and shall be properly identified as required by § 95.95, but may otherwise be unmodulated as appropriate.

§ 95.95 Station identification.

(a) The call sign of a citizens radio station shall consist of three letters followed by four digits.

(b) Each transmission of the station call sign shall be made in the English language by each unit, shall be complete, and each letter and digit shall be separately and distinctly transmitted. Only standard phonetic alphabets, nationally or internationally recognized, may be used in lieu of pronunciation of letters for voice transmission of call signs. A unit designator or special identification may be used in addition to the station call sign but not as a substitute therefor.

(c) Except as provided in paragraph (d) of this section, all transmission from each unit of a citizens radio station shall be identified by the transmission of its assigned call sign at the beginning and end of each transmission or series of transmissions, but at least at intervals not to exceed ten (10) minutes.

(d) Unless specifically required by the station authorization, the transmissions of a citizens radio station need not be identified when the station (1) is a Class A station which automatically retransmits the information received by radio from another station which is properly identified or (2) is not being used for telephony emission.

(e) In lieu of complying with the requirements of paragraph (c) of this section, Class A base stations, fixed stations, and mobile units when communicating with base stations may identify as follows:

(1) Base stations and fixed stations of a Class A radio system shall transmit their call signs at the end of each transmission or exchange of transmissions, or once each 15-minute period of a continuous exchange of communications.

(2) A mobile unit of a Class A station communicating with a base station of a Class A radio system on the same frequency shall transmit once during each exchange of transmissions any unit identifier which is on file in the station records of such base station.

(3) A mobile unit of Class A stations communicating with a base station of a Class A radio system on a different frequency shall transmit its call sign at the end of each transmission or exchange of transmissions, or once each 15-minute period of a continuous exchange of communications.

§ 95.97 Operator license requirements.

(a) No operator license is required for the operation of a citizens radio station except that stations manually transmitting Morse Code shall be operated by the holders of a third or higher class radiotelegraph operator license.

(b) Except as provided in paragraph (c) of this section, all transmitter adjustments or tests while radiating energy during or coincident with the construction, installation, servicing, or maintenance of a radio station in this service, which may affect the proper operation of such stations, shall be made by or under the immediate supervision and responsibility of a person holding a first- or second-class commercial radio operator license, either radiotelephone or radio telegraph, as may be appropriate for the type of emission employed, and such person shall be responsible for the proper functioning of the station equipment at the conclusion of such adjustments or tests. Further, in any case where a transmitter adjustment which may affect the proper operation of the transmitter has been made while not radiating energy by a person not the holder of the required commercial radio operator license or not under the supervision of such licensed operator, other than the factory assembling or repair of equipment, the transmitter shall be checked for compliance with the technical requirements of the rules by a commercial radio operator of the proper grade before it is placed on the air.

(c) Except as provided in § 95.53 and in paragraph (d) of this section, no commercial radio operator license is required to be held by the person performing transmitter adjustments or tests during or coincident with the construction, installation, servicing, or maintenance of Class C transmitters, or Class D transmitters used at stations authorized prior to May 24, 1974: *Provided,* That there is compliance with all of the following conditions:

(1) The transmitting equipment shall be crystal-controlled with a crystal capable of maintaining the station frequency within the prescribed tolerance;

(2) The transmitting equipment either shall have been factory assembled or shall have been provided in kit form by a manufacturer who provided all components together with full and detailed instructions for their assembly by nonfactory personnel;

(3) The frequency determining elements of the transmitter, including the crystal(s) and all other components of the crystal oscillator circuit, shall have been preassembled by the manufacturer, pretuned to a specific available frequency, and sealed by the manufacturer so that replacement of any component or any adjustment which might cause off-frequency operation cannot be made without breaking such seal and thereby voiding the certification of the manufacturer required by this paragraph;

(4) The transmitting equipment shall have been so designed that none of the transmitter adjustments or tests normally performed during or coincident with the installation, servicing, or maintenance of the station, or during the normal rendition of the service of the station, or during the final assembly of kits or partially preassembled units, may reasonably be expected to result in off-frequency operation, excessive input power, overmodulation, or excessive harmonics or other spurious emissions; and

(5) The manufacturer of the transmitting equipment or of the kit from which the transmitting equipment is assembled shall have certified in writing to the purchaser of the equipment (and to the Commission

station in this service, which may affect the proper operation of such stations, shall be made by or under the immediate supervision and responsibility of a person holding a first- or second-class commercial radio operator license, either radiotelephone or radio telegraph, as may be appropriate for the type of emission employed, and such person shall be responsible for the proper functioning of the station equipment at the conclusion of such adjustments or tests. Further, in any case where a transmitter adjustment which may affect the proper operation of the transmitter has been made while not radiating energy by a person not the holder of the required commercial radio operator license or not under the supervision of such licensed operator, other than the factory assembling or repair of equipment, the transmitter shall be checked for compliance with the technical requirements of the rules by a commercial radio operator of the proper grade before it is placed on the air.

§ 95.101 Posting station license and transmitter identification cards or plates.

(a) The current authorization, or a clearly legible photocopy thereof, for each station (including units of a Class C or Class D station) operated at a fixed location shall be posted at a conspicuous place at the principal fixed location from which such station is controlled, and a photocopy of such authorization shall also be posted at all other fixed locations from which the station is controlled. If a photocopy of the authorization is posted at the principal control point, the location of the original shall be stated on that photocopy. In addition, an executed Transmitter Identification Card (FCC Form 452–C) or a plate of metal or other durable substance, legibly indicating the call sign and the licensee's name and address, shall be affixed, readily visible for inspection, to each transmitter operated at a fixed location when such transmitter is not in view of, or is not readily accessible to, the operator of at least one of the locations at which the station authorization or a photocopy thereof is required to be posted.

(b) The current authorization for each station operated as a mobile station shall be retained as a permanent part of the station records, but need not be posted. In addition, an executed Transmitter Identification Card (FCC Form 452–C) or a plate of metal or other durable substance, legibly indicating the call sign and the licensee's name and address, shall be affixed, readily visible for inspection, to each of such transmitters: *Provided,* That, if the transmitter is not in view of the location from which it is controlled, or is not readily accessible for inspection, then such card or plate shall be affixed to the control equipment at the transmitter operating position or posted adjacent thereto.

§ 95.103 Inspection of stations and station records.

All stations and records of stations in the Citizens Radio Service shall be made available for inspection upon the request of an authorized representative of the Commission made to the licensee or to his representative (see § 1.6 of this chapter). Unless otherwise stated in this part, all required station records shall be maintained for a period of at least 1 year.

§ 95.105 Current copy of rules required.

Each licensee in this service shall maintain as a part of his station records a current copy of Part 95, Citizens Radio Service, of this chapter.

§ 95.107 Inspection and maintenance of tower marking and lighting, and associated control equipment.

The licensee of any radio station which has an antenna structure required to be painted and illuminated pursuant to the provisions of section 303(q) of the Communications Act of 1934, as amended, and Part 17 of this chapter, shall perform the inspection and maintain the tower marking and lighting, and associated control equipment, in accordance with the requirements set forth in Part 17 of this chapter.

§ 95.111 Recording of tower light inspections.

When a station in this service has an antenna structure which is required to be illuminated, appropriate entries shall be made in the station records in conformity with the requirements set forth in Part 17 of this chapter.

§ 95.113 Answers to notices of violations.

(a) Any licensee who appears to have violated any provision of the Communications Act or any provision of this chapter shall be served with a written notice calling the facts to his attention and requesting a statement concerning the matter. FCC Form 793 may be used for this purpose.

(b) Within 10 days from receipt of notice or such other period as may be specified, the licensee shall send a written answer, in duplicate, direct to the office of the Commission originating the notice. If an answer cannot be sent nor an acknowledgment made within such period by reason of illness or other unavoidable circumstances, acknowledgment and answer shall be made at the earliest practicable date with a satisfactory explanation of the delay.

(c) The answer to each notice shall be complete in itself and shall not be abbreviated by reference to other communications or answers to other notices. In every instance the answer shall contain a statement of the action taken to correct the condition or omission complained of and to preclude its recurrence. If the notice relates to violations that may be due to the physical or electrical characteristics of transmitting apparatus, the licensee must comply with the provisions of § 95.53, and the answer to the notice shall state fully what steps, if any, have been taken to prevent future violations, and, if any new apparatus is to be installed, the date such apparatus was ordered, the name of the manufacturer, and the promised date of delivery. If the installation of such apparatus requires a construction permit, the file number of the application shall be given, or if a file number has not been assigned by the Commission, such identification shall be given as will permit ready identification of the application. If the notice of violation relates to lack of attention to or improper operation of the transmitter, the name and license number of the operator in charge, if any, shall also be given.

§ 95.115 False signals.

No person shall transmit false or deceptive communications by radio or identify the station he is operating by means of a call sign which has not been assigned to that station.

§ 95.117 Station location.

(a) The specific location of each Class A base station and each Class A fixed station and the specific area of operation of each Class A mobile station shall be indicated in the application for license. An authorization may be granted for the operation of a Class A base station or fixed station in this service at unspecified temporary fixed locations within a specified general area of operation. However, when any unit or units of a base station or fixed station authorized to be operated at temporary locations actually remains or is intended to remain at the same location for a period of over a year, application for separate authorization specifying the fixed location shall be made as soon as possible but not later than 30 days after the expiration of the 1-year period.

(b) A Class A mobile station authorized in this service may be used or operated anywhere in the United States subject to the provisions of paragraph (d) of this section: *Provided*, That when the area of operation is changed for a period exceeding 7 days, the following procedure shall be observed:

(1) When the change of area of operation occurs inside the same Radio District, the Engineer in Charge of the Radio District involved and the Commission's office, Washington, D.C., 20554, shall be notified.

(2) When the station is moved from one Radio District to another, the Engineers in Charge of the two Radio Districts involved and the Commission's office, Washington, D.C., 20554, shall be notified.

(c) A Class C or Class D mobile station may be used or operated anywhere in the United States subject to the provisions of paragraph (d) of this section.

(d) A mobile station authorized in this service may be used or operated on any vessel, aircraft, or vehicle of the United States: *Provided*, That when such vessel, aircraft, or vehicle is outside the territorial limits of the United States, the station, its operation, and its operator shall be subject to the governing provisions of any treaty concerning telecommunications to which the United States is a party, and when within the territorial limits of any foreign country, the station shall be subject also to such laws and regulations of that country as may be applicable.

§ 95.119 Control points, dispatch points, and remote control.

(a) A control point is an operating position which is under the control and supervision of the licensee, at which a person immediately responsible for the proper operation of the transmitter is stationed, and at which adequate means are available to aurally monitor all transmissions and to render the transmitter inoperative. Each Class A base or fixed station shall be provided with a control point, the location of which will be specified in the license. The location of the control point must be the same as the transmitting equipment unless the application includes a request for a different location. Exception to the requirement for a control point may be made by the Commission upon specific request and justification therefor in the case of certain unattended Class A stations employing special emissions pursuant to § 95.47(e). Authority for such exception must be shown on the license.

(b) A dispatch point is any position from which messages may be transmitted under the supervision of the person at a control point who is responsible for the proper operation of the transmitter. No authorization is required to install dispatch points.

(c) Remote control of a Citizens radio station means the control of the transmitting equipment of that station from any place other than the location of the transmitting equipment, except that direct mechanical control or direct electrical control by wired connections of transmitting equipment from some other point on the same premises, craft, or vehicle shall not be considered remote control. A Class A base or fixed station may be authorized to be used or operated by remote control from another fixed location or from mobile units: *Provided,* That adequate means are available to enable the person using or operating the station to render the transmitting equipment inoperative from each remote control position should improper operation occur.

(d) Operation of any Class C or Class D station by remote control is prohibited except remote control by wire upon specific authorization by the Commission when satisfactory need is shown.

§ 95.121 Civil defense communications.

A licensee of a station authorized under this part may use the licensed radio facilities for the transmission of messages relating to civil defense activities in connection with official tests or drills conducted by, or actual emergencies proclaimed by, the civil defense agency having jurisdiction over the area in which the station is located: *Provided,* That:

(a) The operation of the radio station shall be on a voluntary basis.

(b) [Reserved]

(c) Such communications are conducted under the direction of civil defense authorities.

(d) As soon as possible after the beginning of such use, the licensee shall send notice to the Commission in Washington, D.C., and to the Engineer in Charge of the Radio District in which the station is located, stating the nature of the communications being transmitted and the duration of the special use of the station. In addition, the Engineer in Charge shall be notified as soon as possible of any change in the nature of or termination of such use.

(e) In the event such use is to be a series of pre-planned tests or drills of the same or similar nature which are scheduled in advance for specific times or at certain intervals of time, the licensee may send a single notice to the Commission in Washington, D.C., and to the Engineer in Charge of the Radio District in which the station is located, stating the nature of the communications to be transmitted, the duration of each such test, and the times scheduled for such use. Notice shall likewise be given in the event of any change in the nature of or termination of any such series of tests.

(f) The Commission may, at any time, order the discontinuance of such special use of the authorized facilities.

SUBPART E—OPERATION OF CITIZENS RADIO STATIONS IN THE UNITED STATES BY CANADIANS

§ 95.131 Basis, purpose and scope.

(a) The rules in this subpart are based on, and are applicable solely to the agreement (TIAS #6931) between the United States and Canada, effective July 24, 1970, which permits Canadian stations in the General Radio Service to be operated in the United States.

(b) The purpose of this subpart is to implement the agreement (TIAS #6931) between the United States and Canada by prescribing rules under which a Canadian licensee in the General Radio Service may operate his station in the United States.

§ 95.133 Permit required.

Each Canadian licensee in the General Radio Service desiring to operate his radio station in the United States, under the provisions of the agreement (TIAS #6931), must obtain a permit for such operation from the Federal Communications Commission. A permit for such operation shall be issued only to a person holding a valid license in the General Radio Service issued by the appropriate Canadian governmental authority.

§ 95.135 Application for permit.

(a) Application for a permit shall be made on FCC Form 410–B. Form 410–B may be obtained from the Commission's Washington, D.C., office or from any of the Commission's field offices. A separate application form shall be filed for each station or transmitter desired to be operated in the United States.

(b) The application form shall be completed in full in English and signed by the applicant. The application must be filed by mail or in person with the Federal Communications Commission, Gettysburg, Pa. 17325, U.S.A. To allow sufficient time for processing, the application should be filed at least 60 days before the date on which the applicant desires to commence operation.

(c) The Commission, at its discretion, may require the Canadian licensee to give evidence of his knowledge of the Commission's applicable rules and regulations. Also the Commission may require the applicant to furnish any additional information it deems necessary.

§ 95.137 Issuance of permit.

(a) The Commission may issue a permit under such conditions, restrictions and terms as it deems appropriate.

(b) Normally, a permit will be issued to expire 1 year after issuance but in no event after the expiration of the license issued to the Canadian licensee by his government.

(c) If a change in any of the terms of a permit is desired, an application for modification of the permit is required. If operation beyond the expiration date of a permit is desired an application for renewal of the permit is required. Application for modification or for renewal of a permit shall be filed on FCC Form 410–B.

(d) The Commission, in its discretion, may deny any application for a permit under this subpart. If an application is denied, the applicant will be notified by letter. The applicant may, within 30 days of the mailing of such letter, request the Commission to reconsider its action.

§ 95.139 Modification or cancellation of permit.

At any time the Commission may, in its discretion, modify or cancel any permit issued under this subpart. In this event, the permittee will be notified of the Commission's action by letter mailed to his mailing address in the United States and the permittee shall comply immediately. A permittee may, within 30 days of the mailing of such letter, request the Commission to reconsider its action. The filing of a request for reconsideration shall not stay the effectiveness of that action, but the Commission may stay its action on its own motion.

§ 95.141 Possession of permit.

The current permit issued by the Commission, or a photocopy thereof, must be in the possession of the operator or attached to the transmitter. The license issued to the Canadian licensee by his government must also be in his possession while he is in the United States.

§ 95.143 Knowledge of rules required.

Each Canadian permittee, operating under this subpart, shall have read and understood this Part 95, Citizens Radio Service.

§ 95.145 Operating conditions.

(a) The Canadian licensee may not under any circumstances begin operation until he has received a permit issued by the Commission.

(b) Operation of station by a Canadian licensee under a permit issued by the Commission must comply with all of the following:

(1) The provisions of this subpart and of Subparts A through D of this part.

(2) Any further conditions specified on the permit issued by the Commission.

§ 95.147 Station identification.

The Canadian licensee authorized to operate his radio station in the United States under the provisions of this subpart shall identify his station by the call sign issued by the appropriate authority of the government of Canada followed by the station's geographical location in the United States as nearly as possible by city and state.

United States of America
Federal Communications Commission

Form Approved
GAO No. B-180227(R01 02)

FCC FORM 505

August 1975

APPLICATION FOR CLASS C OR D STATION LICENSE IN THE CITIZENS RADIO SERVICE

INSTRUCTIONS

A. Print clearly in capital letters or use a typewriter. Put one letter or number per box. Skip a box where a space would normally appear.

B. Enclose appropriate fee with application. Make check or money order payable to Federal Communications Commission. DO NOT SEND CASH. No fee is required of governmental entities. For additional fee details see FCC Form 76-K, or Subpart G of Part 1 of the FCC Rules and Regulations, or you may call any FCC Field Office.

C. Mail application to Federal Communications Commission, P.O. Box 1010, Gettysburg, Pa. 17325

NOTICE TO INDIVIDUALS REQUIRED BY PRIVACY ACT OF 1974

Sections 301, 303 and 308 of the Communications Act of 1934 and any amendments thereto (licensing powers) authorize the FCC to request the information on this application. The purpose of the information is to determine your eligibility for a license. The information will be used by FCC staff to evaluate the application, to determine station location, to provide information for enforcement and rulemaking proceedings and to maintain a current inventory of licensees. No license can be granted unless all information requested is provided.

1. Complete **ONLY** if license is for an Individual or Individual Doing Business AS

FIRST NAME | INIT | LAST NAME

2. DATE OF BIRTH

MONTH | DAY | YEAR

3. Complete **ONLY** if license is for a business, an organization, or Individual Doing Business AS

NAME OF BUSINESS OR ORGANIZATION

4. Mailing Address

4A. NUMBER AND STREET

NOTE:
Do not operate until you have your own license. Use of any call sign not your own is prohibited

4B. CITY

4C. STATE

4D. ZIP CODE

5. If you gave a P.O. Box No., RFD No., or General Delivery in Item 4A, you must also answer items 5A, 5B, and 5C.

5A. *NUMBER AND STREET WHERE YOU OR YOUR PRINCIPLE STATION CAN BE FOUND (If your location can not be described by number and street, give other description, such as, on RT. 2, 3 mi., north of York.)*

5B. CITY

5C. STATE

6. *Type of Applicant* **(Check Only One Box)**

- ☐ *Individual*
- ☐ *Association*
- ☐ *Corporation*
- ☐ *Business Partnership*
- ☐ *Governmental Entity*
- ☐ *Sole Proprietor or Individual/Doing Business As*
- ☐ *Other (Specify)* ________________

7. *This application is for*

- ☐ *New License*
- ☐ *Renewal*
- ☐ *Increase in Number of Transmitters*

IMPORTANT
Give Official FCC Call Sign

8. *This application is for* **(Check Only One Box)**

- ☐ *Class C Station License (NON-VOICE—REMOTE CONTROL OF MODELS)*
- ☐ *Class D Station License (VOICE)*

9. *Indicate number of transmitters applicant will operate during the five year license period* **(Check Only One Box)**

- ☐ *1 to 5*
- ☐ *6 to 15*
- ☐ *16 or more (Specify No. and attach statement justifying need.)*

10. *Certification* I certify that:

- The applicant is not a foreign government or a representative thereof.
- The applicant has or has ordered a current copy of Part 95 of the Commission's rules governing the Citizens Radio Service. See reverse side for ordering information.
- The applicant will operate his transmitter in full compliance with the applicable law and current rules of the FCC and that his station will not be used for any purpose contrary to Federal, State, or local law or with greater power than authorized.
- The applicant waives any claim against the regulatory power of the United States relative to the use of a particular frequency or the use of the medium of transmission of radio waves because of any such previous use, whether licensed or unlicensed.

WILLFUL FALSE STATEMENTS MADE ON THIS FORM OR ATTACHMENTS ARE PUNISHABLE BY FINE AND IMPRISONMENT. U.S. CODE, TITLE 18, SECTION 1001.

11. ________________

Signature of: Individual applicant, partner, or authorized person on behalf of a governmental entity, or an officer of a corporation or association

12. Date ________________

Sometimes it becomes necessary to return an application. By putting your name and address in the area below, you will enable us to return quickly any application which needs correction or clarification: 1) Put your name on the first line in regular order (for example, Joe Doe); 2) Put your number and street on the second line; 3) Put your city, state, and zip code on the third line.

If necessary, use abbreviations to stay within the guidemarks provided.

ORDERING VOLUME VI OF THE FCC RULES AND REGULATIONS

Part 95, concerning Citizens Radio Service, is contained in Volume VI of the FCC Rules and Regulations. Rules may be ordered from the Superintendent of Documents. Send your name and address along with a check or money order (DO NOT SEND CASH) to the Superintendent of Documents, Government Printing Office, Washington, D.C. 20402, specifying that you want Volume VI of the FCC Rules and Regulations. As of August 1975, the current price is $5.35 for a domestic subscription and $6.70 for a foreign subscription. Prices are subject to change without notice.

☆ U.S. Government Printing Office: 1975—593-955/7

Acknowledgements

This book would not have been possible without the untiring efforts of our editor, Patrick Filley, and the many, many denizens of CB land who have given unsparingly of their time and talent. To those listed below and all who contributed to this exciting project, the authors extend their heartiest thanks:

S.I. Russell, David Reimer, Sylvia Sternstein, Leslie Holzer, Eric Sherman, Stephen Gross, John Sodolski, Vern Newton, Thomas Murphy, Sen. Barry Goldwater, Charles Higgenbotham, John Kearney, David Reimer, Margie Casey, Dick Bleil, C. Phyll Horne, Fred Saunders, James McKinny, Cathy Gurley, Hy Siegel, Virgil Tacey, Chuck Sullivan, Larry Blaustein, Richard Smith, Patty Parr, Dale Ponn, John Trimble, Stephen Krasula, Dugg Collins, Bob Cole, Billy Parker, John Trimble, Charlie Douglas, Bill Mack, Buddy Nichols, Buddy Ray, Beth Sheahan, Tom Kneitel, Leo Sands, Robert Thompson, Gerald Reese, Mike Parkhurst, Don Sears, Bill Fries, Jay Huguely, Kenny Price, Jack Key, Jim Moore, Bill Campeau, Paul Bayer, Bill Bradford, Arthur Stepanian, Larry Blaustein, Adrian Curtis, Ferd Boyce, Michael Greenwald, Joe Getz, Sam Lewis, Peter Schwarz, Martin Roth, Bobbie Ladd, L.G. Goodell, Ron Taylor, J.R. Huffman, William Wanrooy, Peter Guggenheim, Morgan Goodwin, Marilyn Marcus, Ed Falk, Dick Ziff, Bruce Marcus, Nick Leone, Toadfrog, and Paper Doll.